Underground Infrastructures

Underground Infrastructures
Planning, Design, and Construction

R. K. Goel
Chief Scientist
CIMFR Regional Centre
Roorkee, India

Bhawani Singh
Former Professor
IIT Roorkee, India

Jian Zhao
Director, LMR
Ecole Polytechnique Federale de Lausanne
Lausanne, Switzerland

AMSTERDAM • BOSTON • HEIDELBERG • LONDON
NEW YORK • OXFORD • PARIS • SAN DIEGO
SAN FRANCISCO • SINGAPORE • SYDNEY • TOKYO

Butterworth-Heinemann is an imprint of Elsevier

Butterworth-Heinemann is an imprint of Elsevier
225 Wyman Street, Waltham, MA 02451, USA
The Boulevard, Langford Lane, Kidlington, Oxford, OX5 1GB, UK

Library of Congress Cataloging-in-Publication Data
Goel, R. K., 1960-
 Underground infrastructures: planning, design, and construction / R.K. Goel, Bhawani Singh, Jian Zhao.
  p. cm.
 ISBN 978-0-12-397168-5
 1. Underground construction–Planning. 2. Underground construction–Design. 3. Underground storage–Planning.
 4. Underground storage–Design. 5. Underground areas–Design and construction. I. Singh, Bhawani. II. Zhao,
 Jian, 1960- III. Title.
 TA712.G584 2012
 624.1'9–dc23
                                    2011051741

British Library Cataloguing-in-Publication Data
A catalogue record for this book is available from the British Library.

For information on all Butterworth-Heinemann publications
visit our Web site at *http://store.elsevier.com*

Typeset by: diacriTech, Chennai, India

Printed in the United States of America
12 13 14 15 16 17 18   10 9 8 7 6 5 4 3 2 1

Dedicated to readers and researchers

Contents

8. Underground Metro and Road Tunnels

9. Underground Storage of Crude Oil, Liquefied Petroleum Gas, and Natural Gas

10. Civic Facilities Underground

11. Underground Structures for Hydroelectric Projects

12. Underground Shelters for Wartime

13. Underground Storage of Ammunitions and Explosives

14. Underground Nuclear Waste Repositories

15. Contractual Risk Sharing

Preface

Heavenly underground cities are dreams of civil and mining engineers, architects, city planners, and geologists. Everywhere people like underground metros, nearby underground malls, and underground parks in the 21st century. The usefulness of underground infrastructure in business is no less in developed and developing nations. Safety against major devastating earthquakes, landslides, cyclones, and wars is the key attraction in underground cities. Safety costs money. Hence a humble effort has been made to write this comprehensive book on "Underground Infrastructures: Planning, Design, and Construction" for city planners, civil and mining engineers, architects, military engineers, administrators, and municipal authorities. Our dear people may also like reading this simple book. The rail and road tunneling networks in weak hilly areas and mega cities are rightly regarded as engineering marvels by people. The aim of this book is to generate more creative confidence among civil, mining, and nuclear energy engineers; city and town planners; architects; geologists and geophysicists; managers; and administrators.

Earlier Bhawani Singh and R.K. Goel have published three books. The first book is on "Rock Mass Classification—A Practical Approach in Civil Engineering" in 1999. The second book is on "Software for Engineering Control of Landslide and Tunnelling Hazards" in 2002. Subsequently, the third book is on "Tunnelling in Weak Rocks" in 2006. Practicing civil and mining engineers, geologists, geophysicists, and students have enjoyed and used all these books. Everyone has boosted our morale so we have written a fourth book. We pray God that everyone enjoys reading and using this book in imagining real master plans for underground cities or infrastructure. The aforementioned three books are offered for detailed designs of underground opening systems, which are envisaged in their master plans.

The Himalayan region is vast, an amazingly beautiful creation of nature that possesses an extensive rejuvenating life support system. It is also one of the best field laboratories for learning rock mechanics, tunneling, underground space science, engineering geology, and geohazards. The research experience gained in the Himalayas is precious to the whole world.

The authors' foremost wish is to express their deep gratitude to Professor Charles Fairhurst, University of Minnesota; Professor E. Hoek, International Consulting Engineer; Dr. N. Barton, Norway; Professor J.J.K. Daemen, University of Nevada; Professor Ray Sterling, USA; Dr. E. Grimstad, NGI; Professor G.N. Pandey, University of Swansea; Professor John Hudson, UK; Professor J. Nedoma, Academy of Sciences of Czech Republic; Professor V.D. Choubey;

Professor T. Ramamurthy, IITD; Mr. B.B. Deoja, Nepal; Mr. A. Wagner, Switzerland; Professor R.N. Chowdhary, Australia; Professor S. Sakurai, Japan; Professor R. Anbalagan, IITR; Professor M. Kwasniewski, Poland; Dr. B. Singh; Professor B.B. Dhar, Dr. N.M. Raju, late Dr. A.K. Dube, Dr. J.L. Jethwa, Dr. Amalendu Sinha, CIMFR; Dr. V.M. Sharma, ATES; late Professor L.S. Srivastava; Professor Gopal Ranjan, COER; Professor P.K. Jain, IITR; Professor M.N. Viladkar, IITR; Dr. A.K. Dhawan, CSMRS; Dr. V.K. Mehrotra; Dr. H.S. Badrinath; Dr. Prabhat Kumar, CBRI; Dr. P.P. Bahuguna, ISM; Dr. Subhash Mitra, Uttarakhand Irrigation Department; Dr. Rajbal Singh, CSMRS; Professor Mahendra Singh, IITR; Professor N.K. Samadhiya, IITR; Mr. H.S. Niranjan, HBTI; and Dr. Rajesh K. Goel, ONGC, for their constant moral support and vital suggestions and for freely sharing precious field data. The authors are also grateful to the scientists and engineers of CIMFR, CSMRS, IRI Roorkee, IIT Roorkee, IIT Delhi, ATES, AIMIL, HEICO, New Delhi and to all project authorities for supporting field research and sharing data.

The authors are very grateful to their families and friends for their sacrificing spirit. Without their support the writing of this book would have been very difficult.

The authors also thank A.A. Balkema, Netherlands; American Society of Civil Engineers (ASCE), Reston; Ellis Horwood, U.K.; Institution of Mining & Metallurgy, London; John Wiley & Sons, Inc., New York; Springer-Verlag, Germany; Trans Tech., Germany; Wilmington Publishing House, U.K.; Pergamon, Oxford, U.K.; Van Nostrand Reinhold, New York; Wiley Interscience, USA; Elsevier Ltd., U.K.; Bureau of Indian Standards, India; ISO for the kind permission; and all eminent professors, researchers, and scientists whose work is referred to in the book.

All engineers, architects, city planners, and geologists are requested to kindly send their precious suggestions for improving the book to the authors for future editions.

R. K. Goel
Bhawani Singh
Jian Zhao

Chapter 1

Introduction

Life is given to us, we earn it by giving it.

Rabindranath Tagore

1.1 UNDERGROUND SPACE AND ITS REQUIREMENT

The joy of traveling through underground metros, rail, and road tunnels, especially the half-tunnels in mountains and visiting caves, cannot be described. Modern underground infrastructures are really engineering marvels of the 21st century. The space created below the ground surface is generally known as *underground space*. Underground space may either be developed by open excavation in soft strata or soil, the top of which is subsequently covered to get the space below, or created by excavation in hard strata or rock.

Underground space is available almost everywhere, which may provide the site for activities or infrastructure that are difficult or impossible to install aboveground or whose presence aboveground is unacceptable or undesirable. Another fundamental characteristic of underground space lies in the natural protection it offers to whatever is placed underground. This protection is simultaneously mechanical, thermal, acoustic, and hydraulic (i.e., watertight). It is effective not only in relation to the surface, but also within the underground space itself. Thus underground infrastructure offers great safety against all natural disasters and nuclear wars, ultraviolet rays from holes in the ozone layer, global warming, electromagnetic pollution, and massive solar storms.

Increasing population and the developing needs and aspirations of humankind for our living environment require increasing provision of space of all kinds. This has become a high priority for most "mega cities" since the closing years of the 20th century. The world's population is becoming more urbanized, at an unprecedented pace. There were 21 mega cities with populations of more than 10 million people by the year 2000, as predicted earlier; 17 of these cities were in developing countries [1]. There is a need for sustainable development that meets the need of the present generation without compromising the ability of future generations to meet their own needs.

At the same time, growing public concerns for both conservation and quality of life are rightly giving pause to unrestrained development of the cities at ground level. Provision of new urban infrastructure may either coexist or conflict with improvement of the urban environment. In each city, the balance will depend on local priorities and economic circumstances; but unquestionably, environmental considerations are now being accorded greater importance everywhere. Even skyscrapers have three- to four-storied basements.

In the aforementioned scenario, city planners, designers, and engineers have a greater responsibility to foster a better environment for living, working, and leisure activity at the ground surface and are therefore turning increasingly to the creation of space underground to accommodate new transportation, communication and utility networks, and complexes for handling, processing, and storing many kinds of goods and materials. So to stay on top, go underground.

In different countries, various facilities have been built underground. These facilities include:

- Underground parking space
- Rail and road tunnels
- Sewage treatment plants
- Garbage incineration plants
- Underground mass rapid transport systems, popularly known as "underground metro"
- Underground oil storage and supply systems (through pipelines in tunnels)
- Underground cold storage
- Hydroelectric projects with extensive use of underground caverns and tunnels

In Shanghai, China, more than 2 million m^2 of subsurface space has been developed as underground buildings for various uses since modernization in 1980. Underground supermarkets, warehouses, silos, garages, hospitals, markets, restaurants, theaters, hotels, entertainment centers, factories and workshops, culture farms, plantations, subways, and subaqueous tunnels may be found throughout the city of Shanghai [2].

A publication of Royal Swedish Academy of Engineering Sciences "Going Underground" [3] is a useful document describing various uses of underground space. One outstanding example of the use of underground space is an underground ice hockey stadium with a span of 61 m, in Gjovik, Norway, built for the 1994 winter Olympic Games.

"Out of sight, out of mind" summarizes the advantages of creating public awareness of underground space, a remarkable resource that is still largely underdeveloped but available worldwide. The time is ripe for exploring the possibilities of developing underground space for civic utilities in mega cities. It is pertinent to give the example of the Palika Bazar, an underground market in New Delhi, India. As part of the Cannought Place shopping area, the Palika Bazar was built as a cut-and-cover subsurface structure with a beautiful garden created above

it. If the Palika Bazar was on the surface, the garden space would have been lost and surface congestion would have increased immeasurably.

There is a challenge for developing countries because they have to find solutions that are effective, affordable, and locally acceptable, which can be implemented at a rate that keeps pace with the growing problems. So, more rock engineering experts are being consulted now.

1.2 HISTORY OF UNDERGROUND SPACE USE

In the primitive ages, beginning roughly three million years ago, from the time human beings first existed on earth to the Neolithic age of approximately 3000 B.C., underground space was used in the form of cave dwellings so that people could protect themselves from the threats posed by natural (primarily climatic) hazards. The world's biggest cave is 207 m high and 152 m wide in a Vietnam forest. This Hang Son Doong cave is larger than the Dear cave in Sarawak, Malaysia, which is more than 100 m high and 90 m wide. Following this period, in ancient times from roughly 3000 B.C. to A.D. 500, which spanned the civilizations of Egypt, Mesopotamia, Greece, and Rome, technology employed in the construction of tunnels progressed considerably [4].

The earliest examples of underground structures in India were in the form of dwelling pits cut into the compacted loess deposits in Kashmir around 3000 and 500 B.C. This was brought to light by the Archaeological Survey of India (ASI) during excavations in 1960. These pit houses were found to provide excellent protection against cold and severe winter weather as well as the heat of summer. They also offered protection against external attack. Dwellings dating back to 1600 B.C. were also noted at Nagarjuna Konda in Andhra Pradesh state [5].

The world's most beautiful and elaborate rock tunnels, the rock temples in Maharashtra state, cut out of the hardest rock and having a length of some kilometers, indicate early experience in underground engineering by humans. The tunnels of Ellora alone add up to 10.8 km in length. In medieval India, forts and palaces were provided with fountains, underground pathways, basement halls for storage, meeting halls, summer retreats, and water tunnels. Underground constructions in Daulatabad fort, Man Mandir in the palace of King Man Singh, and the 17 basement chambers below the famous TajMahal, are outstanding constructions of medieval kings of India [6].

The mythological story of Ramayana mentions the town of Kishkindha, which was built completely underground and thus enabled its King Bali to win all battles.

Ancient Egyptians gave utmost care to bury their dead in the underground structures, as they believed strongly in life after death. The upper level is in a rectangular shape with a flat top called "mastaba." The lower part is an underground level where the floor was covered by mortar, crushed stones, and straw. Later the idea of mastaba evolved toward the true pyramid. The pyramid contains an underground tomb structure that was built first before construction of the pyramid. These pyramids are as old as 2778 B.C. The great pyramid of Khufo at Giza in Egypt has an incline tunnel as long as 82 m in Giza rock [7].

In the area of the pyramid, ancient Egyptians were fond of underpasses below causeways leading to other pyramids. Sometimes these causeways were considered sacred paths; workers were not permitted to walk on them and therefore it was necessary to cross them by these underpasses. Another example also in the area of the pyramids is the underground drainage canals for the drainage systems in the funerary temples of Khufo and Chephern. In the courtyard of Khufo's temple, a basin of 20 by 30 feet (7 × 10 m) with a depth of approximately 7 feet (2 m) was cut in the rock as a collector tank from which branch canals go out; most of its length was underground. Another example from this area is the underground construction under the Sphinx and the pedestrian tunnel between the pyramids [7].

In June 1992, 5 large man-carved rock caverns were unearthed by four local farmers after they pumped water out of five small pools in their village near the town of Longyou in the Zhejiang Province of China. Subsequently, 19 other caverns were found nearby. These caverns were excavated in Fenghuang Hill, a small hill that is 3 km north in Longyou County. The hill has elevations between 39 and 69 m above mean sea level. The Longyou rock caverns are a group of large ancient underground caverns. They were carved manually in pelitic siltstone in the Quxian Formation of the Upper Cretaceous. They have the following five characteristics: more than 2000 years old, man carved, large spanned, near the ground surface, and medium to hard surrounding rock. This discovery attracted the attention of many specialists from China, Japan, Poland, Singapore, and the United States [8].

There are 23 known large-scale underground cities in the Cappadocia region in Turkey. The underground cities were connected by hidden passages to houses in the region. Hundreds of rooms in the underground cities were connected to each other with long passages and labyrinth-like tunnels. The corridors were made long, low, and narrow to restrict the movement of intruders. Shafts (usually connected with the lowest floor of the underground cities) were used for both ventilation and communication inside the underground cities. Although some researchers claim that the underground settlements were connected to each other with tunnels, conclusive evidence to support this idea has not yet been found [9].

New hydropower projects are being taken up involving construction of more than 1000-km length tunnels with sizes varying from 2.5 to 14 m diameter to add 16,500 MW of hydropower by the end of 11th 5-year plan in India. After the success of the metro rail project in Delhi with state-of-the-art technology, construction of a metro rail project is planned in various cities, including Mumbai, Bangalore, Hyderabad, Lucknow, Pune, Chandigarh, and Howrah-Kolkata. The Indian Railways is constructing the most challenging Jammu–Udhampur–Srinagar–Baramulla railway line in the difficult Himalayan terrain of Jammu and Kashmir State, and there are 42 tunnels with a total length of 107.96 km in the Katra-Quazigund section (142 km). The Konkan Railway Corporation Limited (KRCL) has constructed a 760-km Konkan railway line with 92 tunnels with a total length of 83.6 km. The Border Roads Organisation has planned a prestigious and challenging highway tunnel with a length of 8.9 km under the 3978-m high Rohtang pass on Manali-Leh road, and construction of the tunnel is to start shortly [10]. Six-story underground parks are under construction in New Delhi and are planned in many cities. The government of India is planning a 497-km-long rail link between Bilaspur-Manali and Leh in Himalaya. China is building 5000-km-long rail lines in Himalaya for rapid development there. Interesting case histories of construction are presented extensively in The Master Builder (Vol. 8, No. 8, Sept. 2006, India).

The basic difference in these historical underground spaces and present-day spaces, however, is that those in the past were built or created out of an interest to do something new and creative. However, today underground space is required for sustainable development and for providing better lifestyles to people.

1.3 UNDERGROUND SPACE FOR SUSTAINABLE DEVELOPMENT

When the United Nations was established in 1945, 90% of the world's population lived in rural areas [11]. With time and migration of people toward cities in search of jobs and better lifestyles, the populations of cities grew many fold. Already a billion of these new urban residents live in health- and life-threatening situations, with hundreds of millions living in absolute poverty. At least 250 million people have no access to safe piped water. Four hundred million people lack sanitation.

Ninety percent of the population growth will be in developing countries; 90% of that will be in urban areas. In effect, every year we are witnessing the birth of 20 new cities the size of Washington, DC, or, put another way, more people will be packed into the cities of the developing world in the 21st century than were alive on the entire planet in 1996 [11]. This global trend of migration toward mega cities is the consequence of rapid growth and sometimes destructive wars that face both rural and urban populations the world over.

Most of these huge cities will be located in developing nations with limited financial resources. Will these cities become ecological and human disasters or can they be designed to evolve as healthy, desirable places to live? Planners are beginning to develop planning tools for future mega cities surrounded by sustainable regions, which in turn would be linked to other regions as part of a global economic network. Each region would import resources such as clean water, energy, raw materials, and finished products. Most cities would be surrounded by rural areas that would provide agricultural and other locally generated products. These sustainable regions in turn would export products and services.

To be sustainable, such regions would have to minimize imports, have excellent infrastructure systems, make optimum use of available urban and rural space, and then generate viable exports with an absolute minimum of waste. Even human wastes will have to be treated to high standards, with both waste water and solid wastes being used as recovered resources for the surrounding agricultural lands. What do tunnels and underground space in general have to do with the environment, sustainable development, and an improved habitat? They play an extremely important role now, and we can expect the tunneling industry to take on an even greater role in the future, for example [12],

- Tunnels play a vital environmental role by conveying clean water to urban areas and by conveying waste water out. Most major urban areas depend on tunnels for these services, which function with a minimum of maintenance.
- The usable space of a parcel of land can, in some cases, be almost doubled by adding floor space or bulk storage below the ground surface. Life-cycle cost analysis may reveal underground alternatives to be much more cost-effective.
- It has been demonstrated by several recent earthquakes that tunnels behave very well in earthquakes (<7 M). If urban planners want an important lifeline line to survive an earthquake, they should go out of their way to use tunnels.
- Underground is the only safe location for the storage of nuclear waste and other hazardous or undesirable materials.
- In transit systems, tunnels provide safe, environmentally sound, very fast, and unobtrusive transportation for people in all walks of life in both developed and developing countries.
- Underground space is being used increasingly for industrial, office, and even residential facilities.
- Underground space for bulk storage of food, liquids, and gas has gained increasing acceptance in various areas of the world.
- Congestion and traffic jams in urban areas have been reduced dramatically by the use of underground metros and road tunnel networks.
- In cold regions, the cost of heating underground facilities is reduced significantly. There is no leakage of heat as in superstructures.
- Hygienic conditions, thermal comfort, and silence inside underground facilities will restore inner peace for people after some time.

As engineers, what is our role in this urban future? First and fundamentally, cities are not sustainable without a proper and well-maintained infrastructure. The trend toward concentration of masses of people in mega cities will present tremendous challenges to urban planners, as well as the infrastructure industry. It is important that engineers contribute to livable, functional cities—the ultimate goal of sustainable development strategies. Underground infrastructures are the life support system for mega cities.

More extensive use of the urban underground can help cities reach the goal of sustainable development, but only under the condition of long-term planning.

Although the concept of underground planning is not new, until now it has not been implemented and has not considered all the four resources (space, water, energy, and material) and their interactions. Planning of the urban underground space should be done considering the four resources of the subsurface as the organ of a body that is in a fragile equilibrium. There should be a holistic approach of "multiple-use planning," which considers not only geological and environmental effects but also economic efficiency and social acceptability of underground space development. As a result, cities will be able to make more extensive use of their urban underground without compromising the use of their resources for future generations.

1.4 WHAT SHOULD BE DONE?

The ITA and the international tunneling industry should strongly support Habitat II (Second United Nations Conference on Human Settlements organized in Istanbul, Turkey, in June 1996) and the goals of sustainable development. The leaders of our industry must collaborate with government officials and urban planners to ensure that the planning and construction of infrastructure will be timely and more than adequate to meet the needs for growing future demand. As a result, we must actively design infrastructure not merely for present needs, but for the distant future also.

At the same time, industry leaders must help make the planners and designers of infrastructure fully aware of the phenomenal advantages of underground facilities, especially in terms of improving the environment and in reduced life-cycle costs. We must also demonstrate to planners and government officials that the earlier underground facilities are constructed, the more cost-effective they can be. We should go from whole to parts in planning and not from parts to unknown whole. We should go for the evolving planning. We should have a master plan of underground infrastructure below the mega cities.

Finally, we need to continue to improve and fine-tune our industry so that underground facilities are quicker and less expensive to build, with much less impact on nearby facilities and residents. As such, safe underground facilities can be more attractive and more desirable to both developers and users.

The result will be a strong boost to the tunneling industry in contributing to a better quality of life and a cleaner environment. The world would be a much more difficult and undesirable place to live in without tunnels and the use of underground space in general. Events such as ISRM and ITA congresses and symposia and conferences at local and international levels in other countries provide an impetus for engineers, planners, government representatives, and developers to work together to see that future world development incorporates the required infrastructure and makes the best possible use of the underground resource. The best part is that all of this is not only good for the environment, but will also be good for the underground industry.

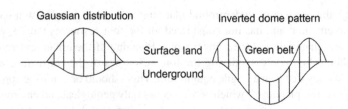

FIGURE 1.1   A Gaussian distribution curve represents a typical urban space use configuration, whereas an inverted dome distribution curve represents the goal of planning of urban space use [4].

The active utilization of land in major cities takes the form of a "Gaussian distribution" curve, as shown in Figure 1.1. The Gaussian curve represents the economic growth of the city, with the center of the city being represented by the central axis. However, in terms of safety and amenities, the greater the tendency of a city toward active utilization of land, the greater the desire to pursue an "inverted dome" pattern of urban planning (Fig. 1.1), which incorporates large-scale public open space. In order to adjust both of these forms to meet cities' needs, increased importance should be placed on the active development of deep underground land as the fundamental starting point for radically reforming the current unattractive, irrational state of surface land development. In short, in order to revitalize cities, in which surface land tends to develop laterally, consideration is now being given to direct the cities' development through the orderly arrangement of a vertical form extending from shallow underground space to deep underground space [4].

1.5 FUTURE OF UNDERGROUND SPACE FACILITIES

It is expected that the cost of excavating underground is likely to be reduced within the next three decades and that the cost of aboveground infrastructure will rise rapidly. As a result, underground space technology may become more economical than superstructures by 2050, in addition to ecological benefits. So, scenic hilly states are likely to grow as future heavens of the globe. Peace-loving civilization is likely to evolve in the near future, as it is need of the time. The new peaceful civilization may opt for healthy technology and ethical economic development of the earth. This peaceful civilization may, therefore, opt for underground space infrastructure to regenerate a healthy ecological heaven on the earth's surface and an underground peaceful heaven with enjoyable homes, facilities, and work culture, providing freedom from mega fears of natural disasters and nuclear wars. Great leaders will continue to be born all over the world in the future who will do great organized works—as in the past so in the future.

A more distant need for underground infrastructure will perhaps occur as part of the drive for developing an occupied base on the moon. A lava tunnel has been discovered near the equator of the moon. Space tourism is likely to become a popular hobby to visit the various planets for sightseeing.

1.6 SCOPE OF THE BOOK

This book is planned to provide a complete knowledge base on various usages of underground space with insight into each type of usage, including its planning, design, advantages, and possible disadvantages. It is planned to give complete information related to underground space use. Interesting case histories of various underground facilities are also given.

One chapter is devoted to each type of underground space use. Important considerations related to the geological and engineering point of view needed while planning are discussed along with the classification of underground space use.

REFERENCES

[1] Weisberg B. Megacity security and social development: a challenge for the world summit for social development. In: Countdown to Istanbul, No. 1; 1995.

[2] Xueyuan H, Yu S. The urban underground space environment and human performance. Tunnel Undergr Space Technol 1988;3(2):193–200.

[3] Winqvist T, Mellgren KE. Going underground. Publication of Royal Swedish Academy of Engineering Sciences. Stockholm 1988. p.177.

[4] Watanabe Y. Deep underground space: the new frontier. Tunnel Undergr Space Technol 1990;5(1–2):9–12.

[5] Goel RK, Dube AK. Status of underground space utilisation and its potential in Delhi. Tunnel Undergr Space Technol 1999;14(3):349–54.

[6] Sharma BD, Selby AR. Subterranean ancient structure of India. Tunnel Undergr Space Technol 1989;4(4):475–9.

[7] Salam MEA. Construction of underground works and tunnels in ancient Egypt. Tunnel Undergr Space Technol 2002;17(3):295–304.

[8] Li LH, Yang ZF, Yue ZQ, Zhang LQ. Engineering geological characteristics, failure modes and protective measures of Longyou rock caverns of 2000 years old. Tunnel Undergr Space Technol 2009;24:190–207.

[9] Aydan O, Ulusay R, Erdogan M. Man-made rock structures in Cappadocia, Turkey and their implications in rock mechanics and rock engineering. ISRM News J 1999;6(1):63–73.

[10] Sharma KG. Numerical analysis of underground structures, IGS annual lecture. Indian Geotech J 2009;39(1):1–63.

[11] Celik AP. The challenges of sustaining our habitat in the twenty-first century. Tunnel Undergr Space Technol 1996;11(4):377–9.

[12] Parker HW. Tunneling, urbanization and sustainable development: the infrastructure connection. Tunnel Undergr Space Technol 1996;11(2):133–4.

Chapter 2

Classification of Underground Space

There are in fact two things, science and opinion; the former begets knowledge, the latter ignorance.

Hippocrates

2.1 GENERAL

This book is written for experts dealing with the design of underground space for various usages. There are a variety of uses of underground space. It is better to group underground space in various classes depending on different parameters, such as position, shape, geometry, and use. Such classification is useful for designers in understanding as to what group/class of underground space is being dealt by them.

The range of potential underground uses is extremely broad even when restricted to potential uses in rock. Underground facilities may include such diverse uses as the use of pore space in rock for fluid or thermal storage, use of natural caves for tourism, reuse of mines for industrial purposes, boring of microtunnels for utilities, and creation of large rock caverns for oil storage or community facilities. This classification is designed to differentiate among types of underground facilities according to factors affecting the choice of an underground facility (i.e., social and geographic aspects, as well as design and construction aspects). Some elements of this classification are presented here to provide a framework for discussion of the various uses and the advantages and disadvantages associated with each type of use. The classification presented here has been mainly taken from [1].

The diverse and overlapping nature of these classification parameters does not permit for a single hierarchical classification scheme. Instead these parameters are most useful in a keyword-type classification scheme, which may allow the grouping of projects in a variety of ways according to any desired combination of classification parameters.

2.2 MAJOR CLASSIFICATION GROUPING

Table 2.1 provides the major classification groupings chosen for underground space use—function, geometry, origin, site features, and project features. Under each category, major subcategories are listed. These subcategories further organize the way in which underground uses are described.

Fenestration in Table 2.1 refers to the classification of underground buildings with a window-like opening arrangement.

2.2.1 Function

Major uses according to their functions are divided into residential, nonresidential, infrastructure, and military uses. Uses are further separated in Table 2.2 into those for which user reactions are important and others for which reactions are either a very secondary consideration or else not applicable. This distinction is considered important as it is easier to develop underground facilities in which user acceptance is not a major issue.

There are few examples of residential structures in rock apart from historical and archaeological uses. Nonresidential underground structures are more common but are more usually built as cut-and-cover buildings in soil than as excavated or mined buildings in rock. Infrastructure uses are the major category of uses for rock excavations, as this category includes uses for mining, energy facilities, utilities, waste disposal, and transportation of both goods and passengers. Military uses are also common in rock and may be separated into operational and civil defense facilities.

TABLE 2.1 Major Classification Groupings of Underground Space [1]

Function	Geometry	Origin	Site Feature	Project Feature
Residential	Type of space	Natural	Geography	Rationale
Nonresidential	Fenestration	Mined	Climate	Design
Infrastructure	Relationship to surface	End use	Land use	Construction
Military	Depth dimensions Scale of project		Ground conditions Building relationships	Age

TABLE 2.2 Classification by Function/Use [1]

Major Function	Subcategories of Use	
	People-Oriented Use	Product-Oriented Use
Residential	Single-family	
	Multiple-family	
Nonresidential	Religious	Industrial
	Recreational	Parking
	Institutional	Storage
	Commercial	Agriculture
Infrastructure	Transportation of passengers	Transportation of goods
		Utilities
		Energy
		Disposal
		Mines
Military	Civil defense	Military facilities

2.2.2 Geometry

Under geometry, classification by type of space recognizes that most underground facilities are made up of relatively few basic geometric elements—pore space/fissures, boreholes/shafts, tunnels, caverns and trenches, or open pit excavations. Because differentiation by fenestration and relationship to surface is typically unimportant for rock structures as compared to earth-sheltered buildings, it is not discussed here.

Classification by depth is always problematical for the wide number of uses and professional backgrounds involved in underground construction. Table 2.3 gives a typical range of depth implied according to various types of uses. The specific dimensions of excavations may also be used to allow the grouping of similar sizes of cavities for comparison of design and performance. The final category—scale of project—allows a differentiation of projects by overall project scale, complexity, and size of investment.

In addition to the aforementioned, the element that characterizes underground structures is the relationship between the structure and the ground surface. Based on this criterion, the uses of underground space may be subdivided into four categories as given in Table 2.4.

TABLE 2.3 Classification of Underground Space Use by Depth [1]

Term	Typical Range of Depth Implied According to Use (m)			
	Local Utilities	Buildings	Regional Utilities/ Urban Transit	Mines
Shallow	0–2	1–10	0–10	0–100
Moderate depth	2–4	10–30	10–50	100–1000
Deep	>4	>30	>50	>1000

TABLE 2.4 Four Categories of Underground Space Use, Based on Relationship between Structure and Ground Surface [2]

Description of Type of Under-Ground Structure	Relationship between Structure and Ground Surface	Main Users	Effects on Aboveground Environment
Located totally underground	Structure is totally below the surface	Shelters, storage, urban traffic facilities, supply management facilities	Preserves open space
Buildings having some floors aboveground and some floors underground	Structure uses both aboveground and belowground space	Offices, pedestrian walkways, parking, warehouses, industry substations	Aboveground buildings can receive more sunlight; deals with height limitations
Atrium-type structures	Structure incorporates atrium(s), skylight(s), to connect the surface with the underground	Pedestrian walkways, residences, sports facilities	Very effective way to preserve scenery and open space aboveground
Underground structures that incorporate shafts	Depends on type of shaft; structures are mainly suited to an inclined plane	Storage facilities, residences	Preserves natural scenery

2.2.3 Origin

Classification by origin differentiates among the use of natural cavities, reuse of excavations left by mining operations, and creation of purpose-excavated cavities for a specific end use. An additional distinction may also be made to cover the adaptation of an existing purpose-excavated cavity for a new use. Reuse of facilities may become more important in the future as the number of existing rock tunnels and caverns increases.

2.2.4 Site Features

Classification by site features may include a wide range of parameters that affect the design or selection of an underground facility—its economic, cultural, topographic, or climatic setting, local land use issues, and local geological conditions. This category is intended to allow the identification of key features that may be useful in understanding the regional or site influences that trigger a particular use of the underground and in grouping projects with similar site conditions.

2.2.5 Project Features

Classification by project feature includes parameters describing the rationale for placing the project underground, major or special design features, and major or special construction techniques utilized.

2.3 BENEFITS AND DRAWBACKS OF UNDERGROUND FACILITIES

Given the wide range of types and sizes of underground facilities, a discussion of the benefits and drawbacks of underground facilities should provide the range of potential impacts on a locational decision of a facility. Table 2.5 gives a listing of issues on the benefits or drawbacks under three categories: physical and institutional issues, life-cycle cost issues, and society issues (which may provide a broad societal benefit or drawback that does not necessarily accrue to an individual project).

A classification of cities based on morphology and soil or rock type is described in Chapter 3.

TABLE 2.5 Benefits and Drawbacks of Underground Facilities [1]

Physical and Institutional Issues		Life-Cycle Cost		Societal Issues	
Potential Benefits	Potential Drawbacks	Potential Benefits	Potential Drawbacks	Potential Benefits	Potential Drawbacks
Location	Location	Initial cost	Initial cost	Land use efficiency	Environmental degradation
Proximity	Unfavorable geology	Land cost saving	Confined work conditions	Transportation/circulation efficiency	Permanent changes
Lack of surface space	Uncertain geology	Construction savings	Ground support	Energy conservation	Embodied energy
Service provision	Isolation	Scale	Limited access	Environment/aesthetics	
Status	Climatic thermal	No structural support	Ground excavation, transportation, and disposal	Disaster readiness	
Isolation	Aesthetic	Weather-independent	Cost uncertainty	National security	
Climatic	Visual impact	Sale of excavated Materials or minerals	Geological	Less construction disruption	
Thermal	Building services	Savings in Specialized design features	Contractual		
Severe weather	Skillful design	Operating cost	Institutional delays		
Fire	Communication	Maintenance	Operating cost		
Earthquake	Environmental	Insurance	Equipment/materials access		
Aesthetic	Site degradation	Energy use	Personnel access		
Visual impact	Drainage		Ventilation		
Interior character	Pollution		Lighting		
Protection	Human issues		Maintenance		
Noise	Psychological acceptability		Repair		
Vibration	Physiological concerns				
Explosion	Fire safety				
Fallout	Personal safety				
Industrial accident	Layout				
Security	Ground support				
Limited access					
Protected surfaces					

Environmental
Natural landscape
Ecology
Runoff
Containment
Hazardous
　materials
Hazardous
　processes
Preservation
Temperature
Conditions
Layout
Topographic
　freedom
Three-dimensional
　planning

Span limitations
Access limitations
Adaptability
Sewage removal
Flooding
Institutional
Easement acquisition
Permits
Building code
Investment
Uncertainty

REFERENCES

[1] Sterling RL, Carmody J. Underground space design. New York: Van Nostrad Reinhold; 1993. p. 328.

[2] Nishi J, Kamo F, Ozawa K. Rational use of urban underground space for surface and subsurface activities in Japan. Tunnel Undergr Space Technol 1990;5(1/2):23–31.

Chapter 3

Important Considerations

We are what and where we are because we have first imagined it.

Donald Curtis

Underground construction differs from other works such as building construction mainly because of its dependence on geological factors, as the rock forms the construction material. The method of construction, duration, and hence final cost depend on the stability and workability of the ground.

Before actually taking up underground planning and construction works, some very important considerations help in making the decision about the use of underground space. These aspects are discussed in the following paragraphs.

3.1 GEOLOGICAL CONSIDERATIONS

Geology is the first basis for land use and town planning. Geology controls the landforms and hence the geometry of any site, from a wide plain to a narrow valley or a mountain slope. Together with climate, geology controls the flow and actions of surficial water, the extent and location of groundwater resources, and the designs for foundations of buildings and other structures. First we should prepare a reliable engineering geological map with several cross sections using extensively the geophysical methods and drilling extensive drill holes. Adequate engineering and geophysical tests must be done in these drill holes. An investment of 2 to 10% should be ensured in geological and geotechnical investigations well before construction of an underground infrastructure. In addition, contingency funds of about 30% need to be arranged for tackling the unforeseen geological risks.

City sites may be classified indicating both the morphology and the rock or soil materials. Duffaut [1] has classified some cities on this basis (Table 3.1). Typically, cities lying directly on bedrock provide better foundations together with better opportunities for the use of underground space. The best example is Kansas City (Missouri) lying upon a flat plateau between the valleys of the Missouri and Kansas rivers. The bedrock is made of horizontal limestone where the mined space has been reused for industrial purposes. However, low plains often provide both poor foundations and bad conditions for underground works due to a groundwater level too close to the surface. The depth to the bedrock also has to be considered. Under the

TABLE 3.1 Classification of Cities by Morphology and Rock or Soil Material [1]

Rock or Soil Type	Morphology				
	Low Plain	High Terrace	Plateau	Rugged Hill	Slope
Sound rock			Kansas City	Oslo	
Soft rock or weathered rock	Lille	Orleans	Madrid	Limoges	Algiers (partly)
Gravels	Strasbourg Lyons (partly)	Toulouse	La Paz (Altiplano)		La Paz (partly)
Fine soils	Chicago	Bangkok	Winnipeg		

Chicago clay lies a sound dolomite where very large tunnels and reservoirs have been bored as part of the "TARP" project to intercept sewer overflows, thus avoiding the pollution of Lake Michigan.

Of course, many cities do not fit any one classification. For example, one part may lie upon a low plain, another over or around some hills. In such a case, the space inside the hills is of special value for city planning, as it is at road level and easy to dewater by gravity.

Beyond a description of subsoil conditions and morphology, local geological features are of great importance for the location, orientation, and design of underground works. For example, in the case of stratified rocks, the orientation and dip of bedding planes are important together with the thickness of the layers. In horizontally bedded rock formations, convenient flat-roofed caverns can be constructed readily with their span controlled by the thickness and competency of the rock beds immediately above the roof.

Geological factors exert an important influence on the design and construction of any underground facility. One of the most important factors is the strength of the soil or rock mass into which underground structures are excavated. Shallow underground structures can be constructed in soil or highly weathered rock using cut-and-cover construction. These structures are built by first excavating a trench or pot, then building the desired facilities, and finally covering the completed structures with soil. Because soil and weathered rock near the earth's surface tend to be weak, shallow cut-and-cover structures must be heavily reinforced with concrete or other materials if they are to withstand attack. The mineral quartz, which can be a common component of the rocks used for concrete aggregate, changes volume when it undergoes a phase transition at high temperatures (844°K at a pressure of 0.1 MPa). In order to prevent thermal disintegration, therefore, aggregate for concrete that may be subjected to extremely high temperatures during fires must consist of rocks containing little quartz (see Section 8.8 on road tunnels).

The primary geological factor controlling underground construction in rock is the nature of the rock mass itself. Strong rock with uniform physical properties is the preferred choice for underground construction. Clandestine tunnels excavated beneath the Korean demilitarized zone by North Korea, for example, tend to be located in granite that is relatively uniform and contains few fractures rather than adjacent rocks that are highly fractured. Depending on the geological setting of an underground facility, selecting the most optimal rock may not be an option. Rocks are commonly heterogeneous, with physical properties such as strength and degree of natural fracturing varying from place to place. Several books on engineering geology and rock mechanics for underground design are available and readers should refer to these books [2,3].

Geoethical questions to be incorporated in all planners', owners', consultants', and contractors' activities can be listed as follows [4]:

- What efforts can be justified to be done before any action in the underground starts?
- Is it enough to safeguard a sustainable solution to the problem?
- How is the control done?
- What are the geodocumentation routines?
- How do we protect the things we all value?
- If it is only a question of cost savings to use the underground, it should also be safe and sound living in the future.
- Why working with geological thinking is one of the fundamental questions to be answered.

It is general thinking that if we could just put problems underground, we could still continue above as before. Geoethics can be said to respect the above and we should all work in line with this. Underground development means that we are utilizing a new space where we have to avoid the same mistakes that we did earlier. We should act, with full respect to existing engineering geology knowledge, now while we still can.

3.2 ENGINEERING CONSIDERATIONS

Successful engineering of tunnels and other underground openings requires an ability to combine a thorough understanding of theory and practice. Underground structures must maintain stability of the surrounding geological environment. In some cases the ground can be self-supporting up to certain span limitations. In cases where support is used, maximum opening sizes are limited by the increasing relative cost of supporting larger openings. Such support costs typically rise more rapidly with clear span than for (surface) superstructures.

Although underground facilities may have a three-dimensional freedom that is not easily possible on the surface, the location of access points for fully underground facilities is limited by surface topography and existing surface uses.

The future expansion or adaptability of underground facilities is also a potential problem. Underground structures are usually expensive to modify and, if designed for a single use, opening sizes and arrangements may not be adaptable for a wide range of other uses.

Topography determines many aspects of what are favorable sites for urban development and national infrastructure. Transportation corridors, land for housing, storage/delivery of water, and removal of sewage are all strongly influenced by topography, especially during the developmental stages of a city. In the same manner, the internal structure of the ground provides natural sites that are more or less favorable to particular types of underground developments.

Surface sites of particular interest (e.g., archaeological, historical, architectural, cultural, or of outstanding natural beauty) are likely to be well defined, documented, protected, and/or regulated. This is seldom true for underground sites with particular characteristics that make them of special interest for historical reasons or for future development.

In relation to surface topology, underground space in hills or near slopes offers many advantages: there is no need to go down to enter, no need for energy to go out, gravity drainage of the structure is possible, etc. Good examples of such spaces are the car parks under the Salzburg Castle hill, the Gjövik sport halls, and, on a larger scale, the Kansas City industrial space within limestone mines.

Engineering geologists and civil engineers commonly describe the physical quality of rock using a simple parameter known as the rock quality designation (RQD), which is obtained by measuring core samples obtained during exploratory drilling prior to construction. The RQD is the percentage of pieces of core sample longer than 10 cm (4 in.) divided by the total length of core. Thus, a core sample of intact rock with no fractures or cracks would have an RQD of 100. A core sample of highly fractured rock in which only one-quarter of the pieces are longer than 10 cm would have an RQD of 25. Other factors that affect the design and construction of underground facilities in rock include the number and density of natural fractures in the rock, the roughness of fracture surfaces and the degree of natural chemical alteration along fracture surfaces (both of which affect rock strength), the presence or absence of water in the fractures, and the presence or absence of zones of weakness such as faults or rock that have been altered to the consistency of clay. Highly fractured rock near a large fault, for example, may be too weak to support itself above an underground cavity or serve as a conduit for high-pressure water that can quickly flood an underground opening [2].

Underground openings in weak, highly fractured, or water-saturated rock may be lined with reinforced concrete or shored with steel beams in order to ensure the safety of construction workers and later occupants of the space. The (lithostatic) in situ stress that must be resisted partly by underground openings of any size increases linearly with depth, and the most stable underground openings are generally circular or spherical. Rectangular or cubic openings contain sharp corners that concentrate stresses in the rock and may lead to cracks in the opening.

Past experience indicates that tunnels are more stable than aboveground structures during an earthquake. However, Dowding [5] concluded that critical frequencies are somewhat lower for caverns than for tunnels because of increased cavern size. Dowding [5] also concluded that cavity response is a function of span or wall height to the earthquake's predominant wave length, dynamic properties of major joints, and relative depth of caverns. Using finite element analysis, Dowding [5] found that for homogeneous rocks, dynamic stress concentrations would be no more than 20% greater than those created by the excavation. Quantitative conclusions could not be verified for "jointed" rock masses.

Yamahara et al. [6] have pointed out that the intensity of earthquake motion at deep base rock may be from one-fourth to one-third of the intensity on the ground surface. They also concluded that rock caverns are extremely safe from earthquakes, given that the rock is practically homogeneous.

This chapter presents in more detail the many advantages of locating facilities underground and also systematically discusses the drawbacks of underground facilities.

3.3 PSYCHOLOGICAL AND PHYSIOLOGICAL CONSIDERATIONS

For common people, the idea of working or living underground elicits a negative (emotional) reaction. Negative associations with underground space generally include darkness combined with humid, stale air, and no sunlight. Among the most powerful associations (emotions) are those related to death and burial or fear of entrapment from structural collapse. Other negative associations (fears) arise in relation to feeling lost or disoriented, as normal reference points such as the ground, sky, sun, and adjacent objects and spaces cannot be seen. Also, with no direct view of the outdoors there may be a loss of connection with the natural world and no stimulation from the variety of changing weather conditions and sunlight. Physiological concerns with the underground focus primarily on the lack of natural light and poor ventilation. People want to see sun and open sky. They do not want to feel like prisoners in underground facilities.

Continuing concern over placing people underground indicates that some of the historic images of dark, damp environments linger in our minds, even though modern technology has overcome many of these concerns. The generally negative reaction to underground space has forced designers and researchers to attempt to overcome these perceptions. Adaptation to live in well-built underground openings takes some time.

There are numerous practical benefits of utilizing the underground for a variety of purposes. For uses such as storage, utility infrastructure, or transportation tunnels, the involvement of people in the space is relatively low and few concerns are raised. However, when underground space is to be utilized for functions that involve human occupancy, initial reactions are often negative and a wide range of concerns and questions are raised. The broad fundamental question is "What are the psychological and physiological effects on people utilizing underground space?" A related question is "If there are negative effects in some cases, what design strategies can be employed to alleviate these concerns and create a positive, healthy environment?"

The purpose of this chapter is to identify potential psychological and physiological problems related to placing people in underground environments. Information is drawn from two sources: (i) the images of the underground that seem to be rooted in history, culture, language, and possibly the subconscious and (ii) the actual experience of people in underground or other analogous enclosed environments. These sources combine to produce a list of potential problems to overcome.

Also noted at the end of the chapter are mitigating factors that influence the impact of these problems, as well as an identification of offsetting positive associations with the underground.

3.3.1 Image of the Underground

Despite the usually well-lighted and well-ventilated examples of modern subsurface environments, the idea of the underground seems to provoke some powerful images and associations from the past. Natural caves that served as shelter to primitive humans are dark, somewhat cold places with humid, stale air. The darkness itself creates a feeling of mystery and fear of the unknown.

Paradoxically, the underground in its role as shelter also evokes more happy emotions with safety, security, and protection. The image of Mother Earth as a source of fertility and life is powerful, yet the space within the earth is usually envisioned as a lifeless and static environment. The mystery of a dark cave inspires fear but also a sense of adventure. These basic emotions are both reflected in and enhanced by the use of underground imagery in literature, religion, language, and psychology.

In order to determine the images associated with the underground, researchers conducted a survey with both Japanese and American subjects. Respondents selected adjectives from a list that conveyed the images of comfort, discomfort, and the underground. The words selected for underground imagery by people in both countries were mostly words that conveyed images of discomfort, although not all words chosen reflected discomfort. Commonly selected words in both groups were fear, uneasiness, and timidity. Americans used anxiety and dejection more often than the Japanese to describe the underground, but they also associated the positive word comfort more often as well. Both Japanese and American subjects used the words expectancy and anticipation to reflect underground imagery, and these words also are associated with comfort. Researchers suggest that designs intended to enhance these more positive associations may be a means of reducing the negative imagery of underground space. They also note that the similarity between Japanese and American responses suggests that design approaches and guidelines developed in one country may be applied in another.

3.3.2 Actual Experiences in Underground Buildings

Two studies of children attending an underground school in New Mexico (the Abo Elementary School) indicated no evidence of greater absenteeism or health problems. In fact, respiratory ailments were reduced because of better control and filtration of the air.

Experience in Europe

Underground factories opened in Sweden in 1946 resulted in negative occupant attitudes as well as frequent reports of headaches and fatigue. A comparison of underground and conventional facilities revealed that underground workers complained more of headaches, fatigue, eye ache, nervousness, and insomnia; however, the incidence of absenteeism was only slightly higher. After becoming accustomed to the underground conditions, absenteeism decreased to the same level as found in the above-grade facility.

While it was concluded that there was no proof of negative physiological effects underground, negative attitudes persisted. In 1958 a follow-up investigation revealed that the negative attitudes associated with working underground had practically disappeared. Moreover, blood tests on 100 workers who had been in the underground facility for 8 years proved normal. According to researchers, however, "the psychological atmosphere remained sensitive."

In another study of Swedish workers in underground factories, there were initial complaints of fatigue, headache, impaired vision, and general depression. When inadequate lighting, ventilation, and inappropriate color schemes were improved, however, the complaints stopped.

In the Scandinavian countries, extensive rock caverns have been excavated in the past 20 years to relieve land use pressures and to provide civil defense facilities. In many cases these caverns are then utilized for community recreational functions such as swimming pools and gymnasiums, as well as art museums and theaters. Office and meeting spaces accompany the other functions and, as noted earlier, there is a history of placing some factories underground.

Experience in the United States

A series of interviews were conducted with employees in underground offices and found that major complaints were of stuffiness and stale air, lack of change and stimulation, and the unnaturalness of being underground all day.

There were other employees who were less negative about working underground and appeared to have accepted their surroundings, but no one was enthusiastic. It should be noted that the subjects interviewed did relatively boring, repetitive work and the underground spaces apparently lacked any significant amenities such as high-quality furnishings, finishes, or lighting.

Workers in a completely underground building had lower levels of satisfaction and higher ratings of anxiety, depression, and hostility than in three other settings—the basement of an above-grade building, an above-grade windowless setting, and an above-grade setting with windows. Adjectives most associated with the underground spaces were "unpleasant, dangerous, musty, dark, smothering, unfriendly, gloomy, and isolating." In contrast, adjectives least associated with the underground were "relaxing, attractive, open, interesting, cheerful, warm, inviting, stimulating, secure, and silent."

Experience in China

In the People's Republic of China, networks of tunnels and caverns have been constructed particularly under major cities for civil defense purposes. In recent years these spaces have been utilized for various manufacturing, storage, and community recreational purposes.

Researchers placed a number of concerns that emerged from the survey into three categories: the entrance, the underground working environment, and concerns for health and safety. Long dark tunnel entrances reinforced negative feelings, and artificial windows to improve the tunnel were disliked. Generally, most people disliked working underground and were unsatisfied with their jobs. They wished for sunlight and greenery and noticed unusual smells underground. Most workers feared illness from the conditions and expressed concerns over their safety in fire, flood, earthquake, or structural collapse. In potential emergencies, they feared losing their way and being unable to escape. Many of these concerns seem to reflect the relatively unsafe and substandard conditions often found in Chinese underground spaces in the past.

Experience in Japan

Severe land use pressures in Japan have led to an intense interest in extensive underground development in recent years. Researchers have begun to conduct surveys to attempt to identify the underlying psychological problems in underground work environments.

There was no significant difference between the two groups with regard to disaster prevention and safety in the underground—both were neutral. With regard to other issues, however, there was a notable difference in the attitudes of above- and below-grade workers toward underground space. More aboveground workers felt that an underground workplace would present a hindrance to achieving a good interior environment and felt they would be extremely burdened psychologically by working underground. A remarkable 60 to 85% of underground workers approved of working below grade, while only 25% of aboveground workers approved. This study reflects the inherently negative attitudes about underground space that are based on images and associations rather than direct experiences.

Another Japanese study seems to be consistent with most of the research in actual underground settings, which indicates predominantly negative feelings about the environment. Workers surveyed in an underground office, subway station, shopping center, and security center reported high levels of anxiety over isolation from the outside (67% of those interviewed) and their physical health (50%). Some anxiety was expressed over mental stress (36%) and earthquake or fire (18%). The most dominant complaints were (in rank order): bad air quality, lack of knowledge about the weather, an oppressive feeling, and low ceilings (even though they were the same as comparable above-grade settings). A few of those interviewed recognized the positive benefits of a quiet environment for work; however, the majority saw no advantages. More than 70% of those interviewed wished to work above (ground) grade, although researchers noted that the physical conditions in fact were not very good in these facilities. It is interesting to note that while these workers were negative about the underground setting, they would continue to work there if conditions in the physical environment were improved, particularly if sunlight, plants, surrogate windows, spaciousness, and more ventilation were provided. They also included a greater willingness to work there if they were compensated financially. A questionnaire on utilization of underground space is given in the Annexure.

3.3.3 Actual Experience in Windowless and Other Analogous Environments

Underground spaces share some basic characteristics with certain analogous environments. These include completely artificial environments such as space capsules, submarines, and arctic or Antarctic bases. Similarities include the almost complete dependence on technology for light and air as well as a sense of enclosure and isolation from the natural environment of the surface. However, these extreme environments present significant differences for people who are confined to them 24 hours a day for periods of a few days to up to several months and in the future, possibly, years. In underground environments, people have periodic access to the surface and at most spend 8 hours a day in the facility. Another key difference is that occupants of these extreme environments are there by choice and are motivated to accomplish a particular mission or scientific endeavor. Because of these differences, overall reactions to these environments do not seem to provide directly analogous information for underground settings.

The lack of windows in buildings above and below grade seems to contribute to the majority of negative attitudes and associations (i.e., claustrophobia, lack of view, natural light, stimulation, and connection to nature). Underground buildings, however, seem to elicit an additional set of negative associations not entirely attributable to lack of windows (i.e., disorientation, coldness, high humidity, poor ventilation, lack of safety, and various cultural and status associations discussed in the previous section). Nevertheless, experience in windowless environments can enhance the overall understanding of people in underground space, but it too is relatively limited.

Windowless Schools

Windowless schools in the United States were constructed to provide emergency shelter as well as to reduce vandalism, glare, overheating, outside distraction, and wasted wall space. In one set of studies in California, there were no significant differences between students in windowed versus windowless classrooms as measured by achievement tests, personality tests, school health records, and grades. A Michigan study revealed similar findings—there were no conclusive detrimental effects on students in windowless classrooms.

Windowless Offices

Ninety percent of female office workers expressed dissatisfaction with the lack of windows, and almost 50% thought the lack of windows affected them or their work negatively. Complaints included no daylight, poor ventilation, inability to know about the weather, inability to see out and have a view, feelings of being "cooped up," isolation and claustrophobia, and feelings of depression and tension. It should be noted that these workers were in small, often single-occupant offices with little freedom of movement.

Windowless Hospital Rooms

Hospital settings have yielded some useful information on the impact of windows as patients are in small spaces continuously and their physiological reactions are being monitored.

Those in a windowless setting developed postoperative delirium twice as often (40% versus 18%) as those in a unit with windows. An increased incidence of postsurgical depression in patients in the windowless setting was also noted. Patients viewing trees had more beneficial results than those viewing a brick wall.

Windowless Factories

Although negative attitudes predominated and there were frequent complaints of physical symptoms, no major physiological problems were discovered.

Functions of Windows

Interesting findings included the fact that people in windowless rooms were less, not more, positive toward windows compared with those in settings with windows. In windowless settings, the trend is for those with the least interesting jobs to miss windows the most. In fact, researchers suggest that windows may be important in reverse proportion to job status. Remarkably, the survey found that blind people missed windows more than sighted people. The sounds of wind, rain, and activity, as well as smells from the outside, were an important source of stimulation for them.

Researchers concluded that windows have multiple functions and effects on the indoor environment. They influence the lighting and thermal environments in several ways; they provide visual and acoustical information from the outside; they affect air quality and ventilation; and they may serve as emergency exits. These functions are described as providing access to environmental information, access to sensory change, a feeling of connection to the outside world, and restoration and recovery.

3.3.4 Potential Psychological Problems Associated with Underground Space

- Because it is largely not visible, an underground building is likely to lack a distinct image.
- Because there is no building mass, finding the entrance can be difficult and confusing.
- Movement at the entrance is usually downward, which potentially elicits negative (emotions) associations and fears.
- Because the overall mass and configuration of the building are not visible and the lack of windows reduces reference points to the exterior, there can be a lack of spatial orientation within underground facilities.
- Because there are no windows, there is a loss of stimulation from and connection to natural and manmade environments on the surface.
- Without windows to the exterior, there can be a sense of confinement or claustrophobia.
- In underground space, there are associations with darkness, coldness, and dampness.
- Underground space sometimes connotes less desirable or lower status space.
- The underground is generally associated with fear of collapse or entrapment in a fire, flood, or earthquake.
- Most artificial lighting lacks the characteristics of sunlight, which raises physiological concerns in environments without any natural light.
- Underground spaces sometimes may have poor ventilation and air quality.
- High levels of humidity, which have potentially negative health effects, are found in underground spaces that are maintained improperly.

Wada and Sakugawa [7] carried out a survey to get the response on various types of anxiety of underground workers while working underground. Results of their survey are given in Figure 3.1.

It is interesting to note that the aforementioned listed potentially negative effects are all related to one of three basic physical characteristics of underground

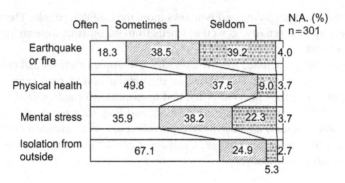

FIGURE 3.1 Various types of anxiety reported by underground workers in response to the question "How often do you feel anxiety at working underground?" [7].

buildings: (1) lack of visibility from the exterior, (2) lack of windows, and (3) being underground. Lack of visibility from the exterior causes the lack of a distinct image and the inability to find the entrance, while it contributes to a lack of spatial orientation inside the building as the overall configuration cannot be understood easily. The absence of windows causes a sense of confinement, lack of stimulation and connection to the outdoors, and lack of sunlight. The windowless nature of underground buildings also contributes to a lack of spatial orientation, as reference points to the exterior are missing, which is related to a fear of not being able to escape in an emergency.

Finally, simply being underground elicits associations with darkness, coldness, dampness, poor air quality, lower social status, and fear of collapse or entrapment.

It is believed that the potentially negative psychological and physiological reactions discussed earlier can be diminished or alleviated to a great extent by utilizing proper design techniques. Although designers of deep, windowless underground space have little specific psychological research to guide them, many of the techniques that follow are based on common sense and have been employed successfully in various underground buildings [8].

3.3.5 Mitigating Factors

- *Building function.* Obviously, all psychological and physiological issues are irrelevant for functions such as utilities and storage underground, whereas they are extremely important for functions that are highly people oriented, such as offices or hospital rooms. Even among the various people-oriented functions, however, there are facility types that are relatively well suited to an enclosed, windowless environment and are often built that way intentionally (i.e., theaters, museums, libraries, gymnasiums, laboratories, and manufacturing plants).

- *Occupancy patterns and freedom of movement.* The effects of an underground environment are mitigated by the amount of time spent there.
- *Type of activity.* Basically, people doing boring, monotonous work seem to complain more about lack of windows than those with more stimulating activities.
- *Social contact and stimulation by internal activity.* The inherent lack of stimulation underground is offset to some degree by social contact and dynamic activity within a space. For example, the windowless nature of a department store does not seem to bother people as much as with other functions due to a continual contact with people as well as constant activity.
- *Size of space.* Smaller spaces such as private offices and hospital or hotel rooms exacerbate the feelings of confinement underground. Larger, more open spaces not only are less claustrophobic but also are likely to contain more activity and stimulation.
- *Degree to which a building is underground.* In deeper facilities, the negative associations with being underground are likely to be greater while the opportunities to provide amenities such as light and view through courtyards are diminished.
- *Quality of interior spaces.* Obviously, the level of furnishings, finishes, lighting, and other amenities will influence perceptions. People living underground find living aboveground painful due to noise pollution.
- *Individual variation.* Most studies of people in windowless underground environments indicate a range of responses. It may be noted that there is no problem of dust (or toxic dust) in underground buildings, which are windowless.

Important strategies for mitigating the physiological and psychological factors are discussed in five general categories [8].

Entrance Design

The entrance to deep underground space may affect its entire image, regardless of the design of the rest of the space. The primary image to avoid is entering through a dark, enclosed area that connotes a basement or a mine. In deeper space, it is impossible to avoid a downward vertical movement; however, the transition from outdoors to deep space can be made a positive rather than negative experience. Specific techniques include the following:

- An entrance that is easily recognizable from the outside (the underground space may be entered through an above-grade building, thereby resolving some transition problems).
- An entrance area that is spacious with high ceilings (entering adjacent to a large interior space that is viewed from the entrance can create a spacious feeling).

- As far as possible, natural light is provided in the entry area (where this is not possible, high levels of artificial light are desirable).
- The entrance should be beautiful to generate emotions of happiness and to attract users.

Sense of Space, View, and Orientation

Deep underground spaces provide virtually no opportunities for typical exterior views. Small enclosed rooms and narrow corridors underground could contribute to many of the negative effects discussed previously, including claustrophobia, lack of orientation, and lack of connection with the outside world. Techniques to alleviate this include the following:

- Glass partitions between spaces, as much as possible
- Ceilings higher than typical dimensions in one-story spaces
- Multilevel spaces whenever possible
- Large central atrium spaces extending from the surface to the deep space
- Optical devices to provide some view of the surface environment

Natural Light

The lack of natural light is often cited as a primary drawback of underground space even though natural light is commonly not provided to extensive areas within above-grade buildings. Three general techniques can be used to provide the benefits of natural light to deep underground space.

- Large, deep atrium spaces extending from the surface to the deep space.
- Various beamed daylight systems utilizing mirrors and lenses to transmit and distribute light from the surface.
- Full-spectrum artificial light. Unfortunately, opportunities to provide natural light in deep space using atriums and beamed day-lighting systems are limited by physical constraints and the available technology to transmit and distribute sunlight through relatively small shafts open to the surface. The use of full-spectrum fluorescent lighting as an alternative can provide some of the physiological benefits associated with sunlight more effectively than other techniques. For example, full-spectrum lights simulate the ultraviolet portion of the spectrum that is normally screened out by window glass. In addition, artificial light—unlike natural light—can be used to provide even light in all spaces at all times. Even where natural light is provided, it can be used in conjunction with full-spectrum artificial light.

Hanamura [9] indicated that light having the same wavelength as that of natural light has been produced for use in conjunction with office, home, and atrium systems to produce the most comfortable environment possible.

Interior Design Elements

The sense of space created by high ceilings and glass partitions, as well as provisions of natural light, provides the basic framework for the interior design elements. Successful interior design underground can result from a variety of design approaches, just as it can in aboveground buildings. Certain elements, however, generally contribute to creating a positive interior environment underground. These include:

- Use of warmer, brighter colors, as opposed to dark colors or completely unfinished surfaces (which connote basement spaces)
- Extensive use of green plants
- Use of water in pools or fountains in appropriate locations
- Variations in lighting in special areas, that is, very bright lights over a plant-filled area or spotlights illuminating artwork
- Artwork that serves as surrogate windows, for example, murals of natural landscapes have been used successfully in windowless spaces to generate emotions of happiness

Mechanical System

Providing fresh, clean air at comfortable temperature and humidity levels should be no more difficult in underground spaces than it is in conventional aboveground buildings. The mechanical system must be designed properly to respond to the unique temperature and humidity conditions that may be found below grade, while also preserving the inherent energy conserving benefits of this isolated environment. Despite the ability to properly design a mechanical system using conventional techniques, the occupants of underground space may be quite sensitive to ventilation, temperature, and humidity problems. To offset negative occupant reactions, strategies include the following:

- Provide ventilation in a manner that is perceptible to the occupants.
- Provide a flexible mechanical system that can control both humidity and temperature to satisfy the function of each space as well as the comfort of the occupants. In other words, underground space should be thermally comfortable.

3.4 CHOOSING TO GO UNDERGROUND—GENERAL ADVANTAGES

Individual projects may only involve a few issues providing significant advantages or disadvantages. It is also important to note that one of the prime considerations in developing underground space is location. Thus, the issue is often not merely choosing between an underground facility and a surface facility on the same site. However, one should also determine whether alternate locations, types of construction, and perhaps alternate means of achieving the desired end result without construction are appropriate.

First, there is usually a problem to be solved, for example, a needed facility expansion or provision of a new service. This can be met by construction of a new facility or service in a particular vicinity but also potentially by reducing the demand for a new facility (using conservation techniques for energy or water facilities, for example). If a new facility is to be considered seriously, then several alternative sites and construction types may be examined with differing attributes in functionality, cost, ease of permitting, and uncertainty. For example, a high-cost option of siting a new surface structure in downtown Tokyo may be compared with an underground structure also in downtown Tokyo or a surface structure in a suburban area. One of the principal issues to be weighed will be the relative importance of a location in the downtown area—a decision unrelated to any specific features of construction.

Direct benefits of a particular project are separated from indirect societal benefits with little relevance to an individual user (unless the environmental benefits are linked to permit costs, tax incentives, etc.). Physical benefits are likewise separated from those benefits that can be expressed in an actual cost benefits to a project. Although some physical benefits may be measured in terms of cost, others, such as aesthetic issues, must be balanced within a decision-making framework.

3.4.1 Potential Physical Benefits

Locational advantages for underground structures include the ability to build in close proximity to existing facilities or on otherwise unbuildable sites. For some projects, such as essential utilities, the location may be predetermined, which in turn may mandate underground construction due to the utility type or surface restrictions. Building decisions may also be affected by the status of certain locations within a city or urban area, which may translate into a premium being paid for a downtown location. For example, a higher cost for underground construction may be acceptable if it permits a downtown location for a facility that otherwise would have to be located further from the city center.

There is an isolational advantage related to the physical characteristics of typical underground spaces and their surrounding ground environment. The ability to isolate structures with a mass of earth provides the following specific advantages.

In most regions of the world the temperature within the soil or rock at depths of less than 500 m represents a moderate thermal environment compared with the extremes of surface temperatures. These moderate temperatures and the slow response of the large thermal mass of the earth provide a wide range of energy conservation and energy storage advantages:

- Conduction losses from the building envelope in cold climates are reduced.
- Heat gain through the exterior envelope from both radiation and conduction is avoided in a hot climate.
- Earth-contact cooling is possible in hot climates.
- Energy requirements are reduced for tempering air infiltration.
- Peak heating and cooling loads are reduced due to large thermal inertia.

Underground structures are protected naturally from hurricanes, tornadoes, thunderstorms, hail, and most other natural phenomena. The most vulnerable portions of underground structures are surface access points for entry, light, or view. Underground structures can also resist structural damage due to floodwater, although special isolation provisions are necessary to prevent inundation of the structure itself.

Underground structures provide a natural protection against external fires. The ground is incombustible and provides excellent thermal isolation to the structure beneath. Access points are again the most vulnerable. Urban fires are a major concern during other calamities such as major earthquakes or during wartime.

Underground structures have several intrinsic advantages in resisting earthquake motions:

- Ground motions on a ground surface are amplified by the presence of surface waves.
- Structures below ground (in rock) are constrained to move with the ground motions so there is not the same opportunity for amplification of ground motions by structural oscillation effects as there is aboveground.

Major earthquakes will damage underground structures near faults; special care must be taken in lightly supported blocky rock structures that could be loosened during ground movements. In general, however, experience with underground structures in rocks during earthquakes has been excellent [2].

3.4.2 Protection

Noise: Small amounts of earth cover are very effective at preventing the transmission of airborne noise. This attribute can be very important for structures located in exceptionally noisy locations, such as those adjacent to freeways and major airports. Surface openings provide the major transmission path for noise to the interior.

Vibration: Major vibration sources in urban areas include road and highway traffic, trains, subways, industrial machinery, and building HVAC systems. High technology manufacturing systems require environments with increasingly stringent limits on vibration amplitudes, velocities, and accelerations. If the vibration sources are at or near the ground surface, levels of vibration will diminish rapidly with depth below ground and distance from the source. High-frequency vibrations are eliminated more quickly with depth than low-frequency vibrations.

Explosion: As with noise and vibration, the earth will absorb the shock and vibrational energy of an explosion. Arching of the soil across even shallowly buried structures greatly increases the peak air pressures that a structure can withstand. Once structural protection has been achieved, access points must be designed to prevent the passage of high air overpressures.

Fallout: Radioactive fallout consists of radioactive dust particles that settle on the ground or on other surfaces following a release of radiation into the air. The majority of the types of radiation present in fallout from an atomic bomb can be absorbed by several inches (centimeters) of concrete, steel, or earth. A major benefit of underground structures in this regard, in addition to their heavy structures and earth cover, is the limited number of openings to the surface. All building openings and the ventilation system must be protected properly to provide adequate fallout protection. A fatal collision of airplanes with buildings is not possible in underground structures.

Industrial accident: Explosion, fallout, and similar catastrophic protection are not solely related to military uses—many industrial facilities have a significant potential for explosions and toxic chemical release. Also, terrorists often strike nonmilitary targets. Underground structures, especially if provided with the ability to exclude or filter contaminated outside air, can be valuable emergency shelter facilities.

3.4.3 Security

Limited access: The principal security advantage for underground facilities is that access points are generally limited and easily secured. This limitation of entry and exit points also seems to inhibit would-be intruders or thieves.

Inaccessibility: The structure of the facility away from the surface opening is generally not directly accessible. Excavation or tunneling for unauthorized access is time-consuming and can be monitored with relative ease.

Containment: Containment is the inverse function of protection. With containment, the goal is to prevent a damaging process to travel from the facility to the surface ecosystem.

Hazardous materials: Underground hazardous material storage can take advantage of the protection, isolation, and security of the facility. Proper design and geological siting can provide very low probabilities of hazardous material leakage and of any such leakage being transported to the surface environment. Hazardous materials include both high- and low-level nuclear wastes and hazardous chemicals that are not disposed of by chemical alteration or incineration.

Hazardous process: Containment of a hazardous (e.g., potentially explosive) process below ground can limit the effect of an explosion on the surrounding community. Several major industrial accidents since 1990 have highlighted this concern, and protective earth berms to deflect blast waves upward are now required for some surface facilities.

3.4.4 Aesthetics

Visual impact: A fully or partially underground structure has less visual impact than an equivalent surface structure. This is important in siting facilities in sensitive locations or when industrial facilities must be sited adjacent to residential areas. The increasing requirement for all utility services to be placed below ground is primarily a visual impact decision.

Interior character: An underground structure can provide an interior character quite different from that of a surface structure. Combinations of tunnels, chambers, and natural rock structures in a quiet, isolated space have inspired many religious expressions. At the other end of the spectrum, the hustle, noises, and smells of a busy subway system also can provide a memorable aesthetic experience.

3.4.5 Environmental Advantages

Natural landscape: This is similar to the limited visual impact mentioned earlier but specifically related to preservation of a natural landscape in keeping with the local environment.

Ecological preservation: When natural vegetation is preserved through the use of underground structures, less damage is inflicted on the local and global ecological cycle. Plant life, animal habitat, and plant transpiration and respiration are maintained to a greater extent than with surface construction. Japan is growing crops, fruits, and herbs successfully in underground openings using artificial light and controlled humidity.

Rainfall retention: Results of the preserved ground surface are that percolation of rainfall to replenish groundwater supplies is encouraged and storm water runoff is reduced. This reduction in runoff permits smaller storm sewers, detention basins, and treatment facilities and also reduces the potential for flooding.

3.4.6 Materials

For example, embalmment followed by entombment has been very successful in preserving corpses from ancient civilizations. Likewise, food preservation is often enhanced by moderate and constant temperature conditions and the ability to maintain a sealed environment that restricts the growth of insect populations and fungi.

3.4.7 Initial Cost/Land Cost Savings

The most likely initial cost saving is the reduced cost for the land purchase necessary to carry out the project. Land or leasement costs for an underground project may range from a full purchase of the site for projects that essentially usurp the surface use to a very low percentage of the land value if no impact on the existing surface use will occur and the surface owner would have little opportunity in developing use at the depth proposed.

3.4.8 Construction Savings

Although underground structures typically cost far more to construct than equivalent surface structures, some combinations of geological environment, scale of facility, and type of facility may provide direct savings in construction cost. An example of this is the Scandinavian experience with large oil storage caverns. Underground structures may also provide weather-independent construction, which can offer some cost advantages in severe cold climates.

3.4.9 Sale of Excavated Material or Minerals

If the underground facility is excavated in a geological material (ore) with an economic value, the sale of this material can be used to offset the excavation cost. If, as in the Kansas City, Missouri, limestone mines, the excavation is part of a profitable mining operation, the space becomes a near-no-cost by-product of the mining. In Coober Pedy, Australia, the rewards are not as certain, but a resident excavating an underground home there can recoup costs if sufficient opals are found during the excavation. In most cases, however, a planned continuous supply of a mineral resource is required before a reasonable economic value may be developed. This limits the economic recovery for isolated, small projects even when the excavated material has value.

3.4.10 Savings in Specialized Design Features

Physical advantages of underground facilities may provide direct cost benefits when compared with a surface facility. For example, thermal isolation may reduce peak load demands for a facility's HVAC system, enabling a smaller, less expensive system to be installed. The same level of security and protection may be available at less cost than that for an aboveground facility. The partial costs for providing space where vibrations are low, there is a, constant temperature, or clean room may also be less underground than at the surface. For buildings that would have an expensive exterior finish aboveground, significant savings can be made below grade where such finishes are unnecessary.

3.4.11 Operating Cost/Maintenance

The physical isolation of underground structures from the environmental effects that deteriorate building components can result in a low maintenance cost for underground structures. The reduced impact of temperature fluctuations, ultraviolet deterioration, freeze–thaw damage, and physical abrasion can reduce the rate of deterioration of underground structures. Other maintenance advantages may include the absence of snow removal problems in underground subways in snowy areas and the slow temperature change in storage facilities when equipment malfunctions. The life span of stable caves is of the order of 1000 years.

3.4.12 Energy Use

The thermal advantages of underground structures usually translate into reduced energy costs to operate them (Fig. 3.2). Although ventilation, drainage, and

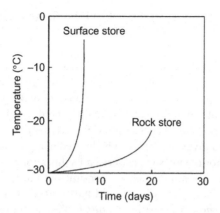

FIGURE 3.2 Rise in temperature after power failure in refrigerated stores [10].

lighting costs often increase, thermal benefits outweigh these in moderate to severe climates, particularly when a surface facility would be designed as a sealed, force-ventilated facility. Savings in energy costs are rarely sufficient to justify building a facility underground for energy conservation reasons alone since there are also techniques for greatly reducing energy use in conventional buildings and the initial cost of an underground building may be high.

3.4.13 Land Use Efficiency

The ability to place support facilities below grade and preserve the land surface for uses requiring the surface environment is an important benefit. With surface development, urban sprawl replaces farmland and recreational areas. Suburban factory development often covers large land areas with windowless buildings and parking lots. The sprawl itself requires more land area to be devoted to automobile and truck transportation because development densities are too low to support adequate urban mass transportation systems. This is being done in some cities by moving railway lines underground and using the reclaimed land for recreational space, infrastructure facilities, or commercial development.

3.4.14 Disaster Readiness/National Security

National safety security concerns provide the driving force for the construction of underground facilities in many rich countries. In Scandinavia, Switzerland, and several other countries, needed community facilities are constructed underground with additional features necessary to provide adequate civil defense shelters. The national government provides for enhanced national security and the community obtains a needed local facility. Underground nuclear shelters are built to protect the lives of leaders. Further, underground bunkers and tunnel networks are made for armies.

3.5 DRAWBACKS OF UNDERGROUND SPACE USE

3.5.1 Physical

Locating facilities below grade requires a greater interaction with the local geological environment than for most surface construction. Sometimes this geological environment may be highly unfavorable for underground construction, and exact geological conditions are difficult to predict prior to construction, which in turn increases project uncertainties. Some types of underground facilities are not restricted to a given area and can be located in a suitable geological environment. However, for others, such as service facilities for existing development, the geological environment in which the construction must take place is already set. Many major urban areas of the world have rather unfavorable geological conditions for underground construction due to their historical development on the estuaries of major river systems.

3.5.2 Climate Isolation

Although mostly a positive issue, the isolation of underground structures may provide thermal disadvantages for certain types of facilities. For example, it is difficult to reject excess heat production in underground facilities except through air conditioning and/or high levels of forced ventilation. Similarly, the slow thermal response of underground structures and cool ground-contact surfaces can cause undesirable interior conditions. In warm, humid weather, a non-air-conditioned building may experience high relative humidities and condensation.

Flooding is a concern for many underground structures, and protection against the effects of surface floods, fire-fighting water, and water leakage from the ground must be provided.

3.5.3 Communication

Communication within underground networks and between the surface and underground spaces may be impeded. Television, radio, GPS, and mobile communication systems will not operate in isolated underground spaces without antennas, cable systems, or special distributed signal repeater systems.

3.5.4 Human Occupancy

Probably the most pervasive drawback to the use of underground facilities for nonservice functions is that a large majority of people express a strong dislike for working in underground or windowless spaces. Coupled with this psychological resistance are concerns for whether an underground environment is healthy for long periods of time in an artificial environment.

3.6 SAFETY CONSIDERATIONS

Safety issues may also represent disadvantages for underground facilities. The ability to exit an underground facility in case of an interior fire or explosion is hampered in deeper underground facilities by limited points of connection to the surface, the need for upward travel on exit stairs, and the difficulty of venting poisonous fumes from a fire. In ground containing dangerous chemicals or gases, these may potentially seep into the underground space, causing health problems. Heavier-than-air gases from the surface may also fall into underground structures to create higher concentrations than would exist on the surface.

As ascertained from Table 3.2, in addition to disasters that attract general attention (such as fires and explosions), disasters that occur with relatively higher frequency include flooding, oxygen shortages, and gas leakages. It should be noted that underground storage facilities (included in the category of "other underground facilities") and basement sections of buildings experience frequent disasters as a result of oxygen shortages and gas leakages. Causes of disasters in underground storage facilities are mostly related to leakage of stored substances, while oxygen shortage and poisoning are the dominant causes of disasters in the basements of buildings [11].

Factors that may increase risks when underground space is developed on a larger scale, with more multiplex uses, and at deeper levels are analyzed next. In addition, safety and disaster prevention measures are suggested for fire/explosion, oxygen shortage/poisoning, flood, and power failure, respectively [11].

3.6.1 Fire Explosion

Measure 1. As a result of the increased complexity of management divisions created by ever more multiplex features in underground spaces, there may be increased difficulty in taking initial action against disasters, for example, through initial fire extinguishing and guidance for evacuation. Countermeasures that may be taken include:

- Unification of fire control districts or closer coordination within each fire control district
- Development of private fire-fighting systems
- Automation of the initial fire extinguishing equipment

Measure 2. Evacuation may become more problematic when the distance for escape is longer. In particular, the escape distance in the vertical direction will increase with development toward deeper levels. Measures to improve safety include:

- Installation of a mechanical evacuation system (lifts) to support upward evacuation
- Securing safety of evacuation routes

TABLE 3.2 Number of Disaster Cases by Subsurface Space Use and by Disaster Category [11][a]

Use of Subsurface Space	Earthquake, Volcanic Activity, and Seismic Sea Waves	Wind and Rain (Excluding Ground Deformation)	Landslide, Landslip, Ground Subsidence, etc.	Snow and Hail	Lightning	Fire (Including Arson)	Explosion and Rupture	Traffic Accident	Structural Failure	Air Pollution, Oxygen Shortage, and Gas Leakage	Crime	Utility Supply Failure	Construction Work-Related Accident	Others	Total
										Causes of Accidents and Disasters					
Ug town	0 (0)	1 (1)	0 (0)	0 (0)	0 (0)	51 (59)	1 (1)	0 (0)	0 (0)	0 (0)	1 (1)	4 (4)	1 (1)	1 (1)	60 (68)
Ug passage	0 (0)	3 (3)	0 (0)	0 (0)	0 (0)	3 (3)	0 (0)	0 (0)	0 (0)	0 (0)	0 (0)	1 (1)	0 (0)	0 (0)	7 (7)
Ug parking lot	0 (0)	3 (3)	0 (0)	0 (0)	0 (0)	6 (7)	0 (0)	2 (2)	0 (0)	4 (4)	1 (6)	0 (0)	1 (1)	1 (1)	18 (24)
Other Ug facilities	0 (0)	0 (0)	0 (0)	0 (0)	0 (0)	2 (5)	3 (9)	0 (0)	0 (0)	21 (27)	0 (0)	0 (0)	1 (2)	1 (1)	28 (44)
Building basement	0 (0)	3 (5)	0 (0)	0 (0)	0 (0)	69 (93)	12 (20)	1 (1)	0 (0)	15 (15)	3 (6)	1 (1)	9 (10)	4 (4)	117 (155)
Mine	0 (0)	0 (0)	3 (3)	0 (0)	0 (0)	8 (9)	11 (12)	0 (0)	8 (8)	9 (9)	0 (0)	1 (1)	0 (0)	3 (3)	43 (45)
Ug station	0 (0)	2 (2)	0 (0)	0 (0)	0 (0)	21 (34)	0 (0)	0 (1)	0 (0)	0 (0)	6 (8)	0 (0)	0 (0)	6 (7)	35 (52)
Ug tunnel	1 (5)	1 (1)	0 (0)	0 (0)	0 (1)	9 (18)	0 (1)	1 (5)	0 (0)	1 (1)	0 (1)	1 (1)	7 (8)	3 (4)	24 (47)
Railroad tunnel	0 (0)	5 (6)	2 (4)	0 (0)	0 (0)	3 (14)	0 (3)	2 (5)	2 (3)	0 (0)	2 (3)	1 (1)	8 (8)	11 (11)	36 (58)
Road tunnel	0 (0)	0 (0)	3 (3)	1 (1)	0 (0)	11 (17)	0 (2)	11 (11)	1 (1)	2 (2)	0 (0)	0 (0)	11 (12)	5 (5)	45 (54)
Other tunnels	0 (0)	0 (0)	0 (0)	0 (0)	0 (0)	0 (0)	0 (0)	0 (0)	0 (0)	0 (1)	0 (0)	0 (1)	5 (8)	0 (0)	5 (10)
Lifelines	2 (2)	7 (7)	6 (6)	1 (1)	1 (1)	8 (11)	8 (22)	5 (6)	0 (0)	70 (79)	4 (6)	1 (1)	57 (64)	37 (37)	207 (243)
Others	0 (0)	0 (0)	0 (0)	0 (0)	0 (0)	0 (0)	0 (0)	0 (0)	0 (0)	0 (0)	0 (0)	0 (0)	1 (1)	0 (0)	1 (2)
TOTAL	3 (7)	25 (28)	14 (16)	2 (2)	1 (2)	191 (270)	35 (71)	22 (32)	11 (12)	122 (138)	17 (31)	10 (11)	101 (115)	72 (74)	626 (809)

[a]Numbers in parentheses are total cases studied; Ug, underground.

- Establishing ultimate safety zones as a final means where safety can be secured even if all of the safety systems fail

Measure 3. Multiplex utilization may make evacuation routes complex, making it difficult for both evacuees and instructors to select the appropriate evacuation routes. Countermeasures include:

- Simple and clear space planning
- Signs (local maps) that enable facility users to recognize their current locations and evacuation routes in an emergency

Measure 4. Enhanced occupancy resulting from a larger space is expected to cause panic in some users. Methods must be developed to permit simultaneous control of numerous evacuees to counter this problem. Nowadays, emergency measures organization in police of high abilities have been organized for fast evacuation of underground metros or in disasters in developed countries.

Measure 5. It is expected that provision of natural ventilation will become more difficult as developments reach deeper levels and multiplex utilization increases. Possible countermeasures include:

- Installation of mechanical smoke ventilation systems that provide sufficient airflow balance
- Installation of open-cut space to facilitate natural ventilation
- Automatic switching on of exhaust fans during fires

Measure 6. Large-scale, deep facilities (particularly in multiplex utilizations) increase the difficulty of identifying fire origin and grasping the actual situation. Disaster prevention measures include:

- Enhancement of monitoring systems by introducing multiple sensors and monitoring cameras
- Installation of automatic fire-extinguishing equipment such as sprinklers

Measure 7. Limitation of approaches to underground space (in particular, to deeper levels) may hamper fire-fighting activities. Establishment of exclusive routes for fire brigades, in addition to evacuation routes, may help alleviate this potential problem. In particular, measures 2 through 4, concerning evacuation, require unified layout and signs in the respective facilities to provide easier and complete recognition by people who may occupy the facilities. These measures are particularly important in consideration of the increased numbers of people occupying the space as the scale of underground projects increases.

A good case history of safety measures against fire taken in Gjovik Olympic Mountain Hall, Norway, is discussed by [12]. A layout of the Olympic hall and other underground facilities is shown in Figure 3.3. The smoke control system is designed to keep the surrounding areas free from smoke by extracting air from the spectator area/arena and supplying air through the tunnels [12].

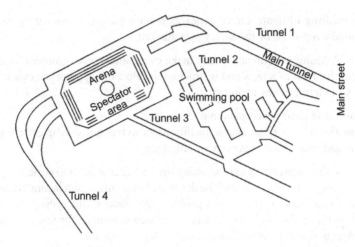

FIGURE 3.3   Gjovik Olympic Mountain Hall. The arena and the spectator area are normally occupied areas. If a fire should occur, it is assumed to have started in one of these areas. In the surrounding area, including tunnels 1 through 4, the extent of combustible and hazardous materials is highly restricted [9].

3.6.2 Oxygen Shortage/Poisoning

Development at deeper levels may limit airflow from the outside, adding to the difficulty of providing natural ventilation. Also, an increase in the number of people occupying an underground space (expanded by development to a larger scale) may lead to greater risks of disasters related to oxygen shortage. Measures to counter this problem include:

- Automatic around-the-clock monitoring of carbon dioxide concentrations.
- Controlling the ventilation volume automatically according to the temporary utilization of space.
- At least 20% oxygen should be maintained in an underground environment.

As a result, oxygen inhalers may be provided at a number of places if one is fainting due to lack of oxygen. Oxygen restaurants may also be provided for enjoying life in underground space.

The lower level of natural ventilation that is likely to result from deeper and larger underground facilities may cause an increase in the exhaust of poisonous and/or explosive gases. Possible countermeasures include:

- Enhancing the systems for detecting and shutting off leakage gases.
- Improving the capacity and reliability of the ventilation systems.
- Providing gas-exhaust facilities with explosion-proof reliability.

3.6.3 Flood

With regard to flooding, the two greatest differences between aboveground facilities and underground facilities are that with underground facilities

(1) there is a greater risk of flooding from above (ground surface) and (2) water has to drain upward. As underground spaces become larger and deeper, the following additional problems with flooding may arise.

- Natural drainage may become more difficult for developments at deeper levels. To combat this potential problem, the capacity and reliability of the drainage systems must be improved.
- Flooding may affect unexpectedly large areas of larger scale developments and multiple-use facilities. One possible solution is the establishment of a waterproof compartment to limit flood damage.
- Failure of groundwater pumping systems may lead to flooding of ordinary spring water. In particular, it is anticipated that flooding and water leakage from upper floors may damage power supply facilities, resulting in power failure accidents. Countermeasures include:
  - A complete waterproofing finish for central sections of the building safety centers, power supply chambers, machine rooms, and computer rooms, as well as enhancement of other measures against water leakage, such as installation of water leakage sensors.
  - Improving the reliability of the power supply facilities and drainage systems.
  - Establishing measures for water conveyance on the respective floors and installing water reservoir pits with appropriate capacities.

3.6.4 Electric Power Failure

There is a great difference in brightness between ordinary lighting and emergency lighting when ordinary lights go off and only emergency lighting is available. It is anticipated that this change in lighting is likely to lead to panic among some people who are negatively affected psychologically. Measures to counter this possibility include:

- Securing illumination by natural lighting if possible.
- Increasing the illuminance of emergency lighting (e.g., compact fluorescent lamp).
- Providing backup measures for a stable power supply in an emergency. It may be necessary to consider backup measures utilizing different sources of energy, such as compressed air, as an ultimate means of securing power.

Underground space development on a larger scale may mean that failure at one facility will affect a number of other facilities. In such a case, any power failure would affect larger areas, and it would be time-consuming work to determine the causes of failure and correct the problem. Measures to alleviate such problems include:

- Distribution of mutual support among power supply facilities.
- Establishment of troubleshooting and diagnostic systems for power supply facilities.

Safety from personal attack can also be diminished in underground structures if public areas are isolated and provide areas for attackers to wait and act unseen. This has been a substantial drawback to some underground pedestrian connections in urban areas. Revolving safety issues require careful design and building operation, which translates into higher costs. The removal of all psychological resistance to underground facilities is not feasible despite the major improvements possible with careful design.

3.7 VENTILATION

In the construction of enclosed and half-enclosed underground structures such as tunnels, underground car parks, and, sometimes, mines, the design of a ventilation system poses a great challenge. Mechanical ventilation systems are installed to supply fresh air and to convey polluted air away from the facility. In excavation of tunnels and exploitation of ore in mines, the air is polluted for two reasons. The first is that carbon monoxide and carbon dioxide are exhausted from machines or vehicles with internal combustion engines. The second is that rock and soil contain gases such as methane and carbon dioxide. When mixed with air, concentrations of methane may be explosive; therefore, this gas must be conveyed away from the area of excavation [13].

In general, a ventilation system is designed by taking into consideration the criteria of maximum and average concentrations of gases in enclosed and half-enclosed underground structures. The design of an appropriate ventilation system for underground structures, in general, must take into consideration, mainly, two issues:

- The supply of the necessary air flow rate that provides adequate ventilation under normal conditions.
- Smoke and high temperature management in case of a fire emergency to establish safe evacuation conditions. According to the methodology presented, the design steps of the ventilation system, for both normal and emergency operation, are:
  - Estimation of the required quantity of the incoming air under normal operating conditions.
  - Determination of the critical air velocity for each branch to control smoke propagation in case of fire. Air velocity lower than the critical can be considered inadequate, whereas larger than the critical velocity could result in a possible increase of the fire size and contribute to its rapid spread throughout the area.
  - Estimation of the required quantity of the total incoming air in order to attain critical velocity and avoid "back layering."
  - Configuration of the facility layout alternatives so as to minimize air quantity requirements.
  - Reestimation of the required air quantity for each alternative.
  - Selection of the final scheme.

A perfect ventilation design is a must for underground space use as per requirements. Hence, it is suggested to consult experts and literature on the subject.

3.8 LEGAL AND ADMINISTRATIVE CONSIDERATIONS

Legal and administrative restrictions on the development and use of underground space may act as significant barriers to the use of this resource. The protection of the rights of existing surface or underground users, the administrative control of mineral reserves of national importance, and the provision of personal safety and environmental protection are issues that must be resolved in all countries [14]. *Municipal bylaws are needed for the use of underground space in all nations.*

3.8.1 Limits of Surface Property Ownership

Because national territories, local jurisdictions, and private ownership are normally defined in terms of boundaries of surface land area, it is necessary for both underground space use and use of airspaces to define how surface ownership extends downward to the underground and upward to the sky.

The most common maxim applied to this definition has been: "The owner of the surface also owns to the sky and to the depths." This extension of surface ownership was common in British, French, Germanic, Jewish, and Roman law and was cited as early as 1250 A.D.

It is possible to legally separate the ownership of the subsurface from that of the surface at particular depths or well-defined changes in geological formation. It is also common to separate the ownership of minerals from surface ownership. When the ownership of underground openings is being defined, it must also be considered that an underground opening depends on a certain volume of soil or rock surrounding the opening for its stability or the overall stability of the ground structure. Thus, it is not likely that two major underground openings can be excavated immediately adjacent but on opposite sides of the same property line.

3.8.2 Ownership and the Right to Develop Subsurface Space

In countries where mineral rights are held by the state, permission must be given to a private company or to a state agency to develop the resource.

Requirements for Permits

Underground excavation and resource removal are regulated to control undesirable environmental impacts and excessive depletion of important natural resources. They are also regulated to ensure safety for existing surface uses and life safety during construction and to provide a healthy and safe environment in an eventual use of the space.

Underground excavations for nonmining purposes often have not had a well-defined regulatory basis because historically most underground excavations were for mining, defense, or utility purposes, not for occupation and use by the public or a regular workforce. Building and life safety codes for surface structures usually have poor application to deep underground structures and mining, and construction regulations do not address the eventual use of space created by the excavation process.

3.8.3 Application of Surface Land Use Regulations

Most countries and regions exert some form of planning control over use of the land surface. This may involve protecting areas of natural beauty, maintaining agricultural land, or setting zoning regulations that control the type and density of development in built-up areas.

Because underground space may allow functions to occur within the space independent of the surface land above, questions often arise as to what extent surface land use regulations should apply to the development of underground space.

Land use regulations may have several aims, for example, to preserve the aesthetic character of an area, to separate incompatible land uses, to avoid overloading community services with new development, or to encourage high-density development in certain areas. It is clear that some of the problems cited are not an issue if a structure is completely underground (the aesthetic issue, for example), whereas others are not solved simply because the structure is underground (overloading the community infrastructure, for example).

3.8.4 Environmental Controls

The environmental impacts of underground developments may include some of the following problems:

- Lowering of regional or local groundwater tables due to pumping or drainage into underground structures. This may in turn lead to the settlement of surface structures and the deterioration of existing building foundations.
- The potential for pollution of groundwater systems from the underground facility.
- The provision of unwanted connections between different aquifers in a regional hydrologic system.
- Disposal of the excavated material.
- Introduction of ground vibrations, for example, subway systems.
- The usual environmental impacts of the type of facility constructed.

3.8.5 Restrictions due to Surface and Subsurface Structures

While mining, tunneling, or creating deep open-pit surface excavations, there is always a potential danger of creating undesirable ground movements that may damage existing surface or underground structures. There are two parts to this issue [15]:

- Is construction of the underground or surface structure permitted if a significant risk exists or if conditions are especially critical (such as damage to a national monument or if a large loss of life could be involved)?
- If it is permitted, who bears the liability for damage to existing structures? This question is not always clear-cut, for example, when development occurs above a mining area with anticipated ground settlement or if the foundations of existing buildings were inadequate and caused damage prior to the effect of any underground structure.

3.9 ECONOMIC CONSIDERATIONS

When considering subsurface facilities, it is possible to distinguish between two general types of installations: compact and (mostly) commercial facilities, such as oil or food storage facilities, and spread-out public service installations, such as public transportation and public utility facilities. Facilities such as those for the underground storage of oil or food can be well defined in a cost–benefit perspective when construction and operation costs are considered. A considerable amount of data is available, and direct cost comparisons between surface and underground installations may be made [16].

With regard to infrastructure facilities, however, a very complex economic situation exists. A common practice in public works planning is to select one alternative from among many on the basis of financial feasibility. The development of subsurface space often is dismissed at this stage because of the lack of comprehensive studies documenting the costs and benefits of underground installations. This, in turn, is due to the fact that a large number of unknown, site-specific factors in subsurface construction make the compilation of required data more difficult than for surface construction.

Promising attempts are being made to provide accurate cost–benefit information for planning considerations. It is clear that many benefits available through subsurface installation cannot be quantified easily and that it remains difficult to assign a value to them.

Unfortunately, underground construction will always be an activity that involves a high amount of risk. Construction cost estimates for underground facilities are more open to discussion and opinion in the early planning stages than construction estimates for surface facilities. This assessment of risk is, in practical terms, very much an economic factor. For example, it is not reasonable for a contractor to assume all the risk because it is difficult to completely know the geotechnical aspects of the project in advance. However, it is not practical to assign all the risk to owners because some judgment (calls) must be made by contractors based on their qualifications and experience.

Before actually taking up the construction of an underground opening for any purpose whatsoever, proper planning is required (see Chapter 4).

REFERENCES

[1] Duffaut P. Subsurface town planning and underground engineering with reference to geology and working methods. In: Proceedings regional symposium on underground works and special foundations, Wiley, Singapore; 1982.

[2] Singh B, Goel RK. Tunnelling in weak rocks. UK: Elsevier Ltd.; 2006. p. 489.

[3] Singh B, Goel RK. Engineering rock mass classification: tunnelling, foundations and land-slides. USA: Elsevier Science Ltd.; 2011.

[4] Morfeldt D. Geo-ethics in underground development. In: Proceedings int conf on underground construction in modern infrastructure, June, Switzerland. Netherlands: A.A. Balkema; 1998. p. 143–50.

[5] Dowding CH. Seismic stability of underground openings. In: Proceedings of the international symposium on storage in excavated rock caverns, Rockstore 77, vol. 2. Oxford, England: Pergamon Press; 1977. p. 231–8.

[6] Yamahara H, Hisatomi Y, Morie T. A study of the earthquake safety of rock caverns. In: Proceedings of the international symposium on storage in excavated rock caverns, Rockstore 77, vol. 2. Oxford, England: Pergamon Press; 1977. p. 377–82.

[7] Wada Y, Sakugawa H. Psychological effects of working underground. Tunnel Undergr Space Technol 1990;5(1/2):33–37.

[8] Carmody JC, Sterling RL. Design strategies to alleviate negative psychological and physiological effects in underground space. Tunnel Undergr Space Technol 1987;2(1):59–67.

[9] Hanamura T. Japan's new frontier strategy: underground space development. Tunnel Undergr Space Technol 1990;5(2):15–21.

[10] Ikiiheimonen P, Leinonen J, Marjosalmi J, Paavola P, Saari K, Salonen A, et al. Underground storage facilities in Finland. Tunnel Undergr Space Technol 1989;4(1):11–15.

[11] Watanabe I, Ueno S, Koga M, Muramoto K, Abe T, Goto T. Safety and disaster prevention measures for underground space: an analysis of disaster cases. Tunnel Undergr Space Technol 1992;7(4):317–24.

[12] Meland O, Lindtorp S. Fire safety and escape strategies for a rock cavern stadium. Tunnel Undergr Space Technol 1994;9(1):31–35.

[13] Likar J, Cadez J. Ventilation design of enclosed underground structures. Tunnel Undergr Space Technol 2000;15(4):477–80.

[14] Esaki T. Underground space and design, chapters 5, 6, Japan. Available from: http://www.ies.kyushu-u.ac.jp/~geo/lecture/underground2005/; 2005. Accessed in 2008.

[15] Stefopoulos EK, Damigos DG. Design of emergency ventilation system for an underground storage facility. Tunnel Undergr Space Technol 2007;22:293–302.

[16] Bergman SM. The development and utilization of subsurface space, U.N. progress report. Tunnel Undergr Space Technol 1986;1(2):115–44.

Chapter 4

Underground Space Planning

It is not the strongest species that survive, nor the most intelligent that survives. It is the one that is the most adaptable to change.

Charles Darwin

4.1 FORMS OF UNDERGROUND SPACE AVAILABLE AND USES

Underground development may be divided into three main areas: underground resource development (e.g., mining), underground energy development, and underground space development. As the amount of ground surface available for development shrinks, underground space should be developed in a variety of ways: to enhance urban life quality, for consumer supply management and production, as storage space, for traffic diversion, and for disaster prevention. Development of diverse types of underground space should consider the proper use of such space for human beings.

The following five main concepts reflect current attitudes toward the potential modern use of underground space.

1. Facilities should be built underground if they do not need to be above-ground; available surface space should be used to beautify cities and enrich the ecosystem.
2. Systems for traffic, communications, energy, and waste disposal should be placed underground to avoid overcrowded areas in cities and inconvenient locations.
3. Facilities for industry, business offices, and recreation may be placed underground.
4. Underground space should be developed in cities subject to severe weather conditions.
5. Underground space should be used for disaster prevention and protection against natural disasters.

The benefits of underground space include a constant and comfortable temperature, constant humidity, heat retention, shielding from natural light, physical and acoustical isolation for noiseless life, air tightness, and safety. These features often lead people to associate underground space with a

Underground Infrastructures

consciousness and mode of life that is quite different from the characteristics associated with the surface (see Section 3.3 in Chapter 3).

Examples of areas most likely to seek underground utilization extending to deep underground space are:

- Cold-climatic regions
- Regions where land prices are high
- Hilly regions that are very spacious

Underground space is intrinsically effective in cities located in cold-climatic regions. For example, cities in northern Europe and North America have played a leading role in underground space utilization. These cities use a diversity of underground utilization forms, such as house basements, shopping centers, subways, roadway tunnels, utility tunnels, parking lots, district heating systems, sewer processing stations, radioactive product storage, disaster shelters, earth-covered houses, and buildings of educational institutions. In short, it can be said that underground space has come to exist as a kind of oasis in these locations.

In hilly regions that have space resources, there is value not only in the rock bed and stone materials that generate mineral resources, but also in the space that is left over after the mineral resources have been extracted. For example, in the Ohya district of Utsunomiya, Japan, Ohya stone is quarried on a regular basis. The space that results in areas where quarrying has been completed is being considered for comprehensive use in the form of production-based factories (e.g., for vegetables, mushrooms, fermented foods); storage-based facilities (for cheese, wine, ham, etc.); prevention-based facilities (e.g., automated three-dimensional warehouses); business-based facilities (e.g., computer centers); culture-based facilities, such as industrial halls and art museums; and leisure facilities (e.g., artificial ski slopes, amusement parks, concert halls, skating rinks). In short, the use of underground space is expected to be able to contribute to the activation of regional core cities with research and development functions, both at present and in the future.

The importance of the use of underground space and functions devoted to this space vary significantly from one country to another and from one city to another. They depend on the historical, practical, and institutional conditions of the development; these are illustrated by the following three examples.

1. Montreal's indoor city was mainly developed because of conjectural conditions. The first opening in the Montreal underground was created for a transcontinental railway (1912 to 1918) in what was then the proposed new Montreal center (which was shifted to reduce the congestion in the old center). To create the opening, the Canadian Northern Railroads had bought more land than what was necessary. When the railway opened in 1943, it rapidly began a major role in the city. The fact that the land was property of a single stakeholder provided an opportunity to plan city development at a large scale; underground development was also favored by climatic conditions [1,2,3].

2. In Tokyo, development of the urban underground started after the major earthquake of 1923 with development of the first metropolitan underground railway. Due to high congestion and a shortage of land, the underground has been developed extensively with rail and pedestrian ways, like in Montreal, but also with major facility infrastructure with roads and tunnels, as well as cultural, energy, and commercial infrastructure. This major development caused congestion within the shallow underground and has given birth to a change in the land property law with a restriction of private property up to a depth of 40 m only [1,4,5].

3. Helsinki has also experienced major underground development. Because the underground infrastructure was favored by good rock conditions in the shallow underground, it has given birth to a multitude of cavern excavations with major architectural realizations such as swimming pools or churches [6,7].

The depth at which underground space is constructed constitutes another complicating factor. Underground space is connected to the ground level through horizontal connections (normally two to three) and vertical shafts (lifts and staircases).

A rock engineering project is always unique. The developed space can be extended, upgraded, and repaired, but it cannot be rehabilitated (although it can, in principle, be filled). Rock caverns are built to last forever virtually, which is an important planning principle. As a result, rock caverns should lend themselves to easy modification, extension, maintenance, and repair, even in the event of future unpredicted changes in use of the space.

4.2 LEVEL-WISE PLANNING OF UNDERGROUND SPACE USE

Underground space can be divided into different categories according to its end use: space intended for use by the general public, traffic space, technical maintenance facilities, industrial and production facilities, and special-use facilities. Figure 4.1 illustrates the feasible depths of different activities in the urban structure.

The first category of space for the general public refers to underground space used for leisure and recreational purposes, exhibition and culture facilities, and shops and service facilities. In the planning of underground space for such uses, special attention must be given to lighting, ventilation, acoustics, ease of orientation and movement, and in general to creating a healthy environment in which people can feel comfortable. The building of underground office space must be restricted to the absolute minimum. Special planning considerations include the amount and quality of lighting and the ventilation system.

Underground traffic facilities refer to different kinds of tunnels and car park facilities. The planning and characteristics of traffic tunnels depend

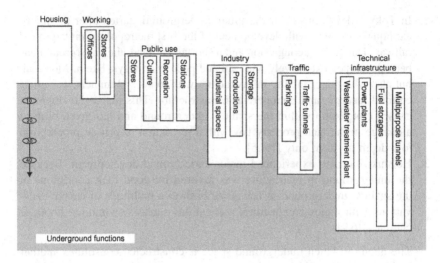

FIGURE 4.1   Feasible depths of different activities in urban structures [6].

significantly on their specific use, that is, whether they are intended for motor traffic, maintenance, light traffic, or rail traffic (metro and railway). Technical maintenance facilities comprise underground space for sewage treatment plants, power plants, fuel supplies, depots, and fire and rescue service facilities. In addition, energy management and municipal infrastructure tunnels are included in this category. Industrial and production facilities comprise industrial manufacturing, production, and storage space, as well as maintenance and repair facilities. Special-use facilities include equipment shelters, telecommunications facilities, and control and command centers, as well as research, testing, and laboratory facilities.

In a nutshell, moving from shallow layers to deeper layers, the layers of underground space that can be used, comprise Levels 4 to 1. Facilities on Level 4 – offices, exhibitions, commercial facilities, and recreation halls – are used routinely by a large number of people. Level 3 facilities, which include transportation facilities and underground parking lots, are used for a fixed period of time by an unspecified large number of people. Level 2 comprises facilities used only by trained personnel – power plants, transformer stations, production facilities, etc. Unmanned facilities – waterworks and sewer lines, gas lines, telephone lines, common ducts, underground waterways, etc. – are constructed in Level 1, the deepest level.

The space use arrangements just described represent optimal uses of the various layers of underground space. In practice, however, the shallow layers of underground space exist on a "first come, first served" basis; the sequence of use described frequently differs from the actual case. In addition, the way in which the four levels are combined is very important. At the very least, underground structures should be connected for operation and control and should provide access to the surface. Because access ways to the surface, in addition to regulating the flow of people and objects underground, serve to deliver elements essential to living things, such as light and air, it is important to build underground structures that are capable of fulfilling these roles adequately.

What is now being sought from deep underground utilization (taking a more challenging view with respect to it) is the creation of an underground environment that goes beyond Levels 3 and 4 and extends to Levels 1 and 2. The technology that will make this goal a reality should be pursued. In terms of urban form, the inverted dome (arch) configuration has been evaluated as being more beneficial to human beings from the viewpoint of the safety and social amenities it provides, although this urban form will need to be adjusted to take economic characteristics into consideration under a considerably long-term planned development and utilization project. As a result, the position and suitable utilization of underground space within the urban form are important [8].

4.3 FUTURE FORMS OF UNDERGROUND SPACE USE

Large-scale redevelopment, which is indispensable to urban planning, as well as the use of underground space in the form of new, attractive urban areas in harmony with nature, is likely to appear in the future.

Redevelopment planning may achieve a combination of underground space and ground surface space involving the use of Levels 1–4 described previously.

In terms of revenues, it is believed that it will be possible to use prepaid rent from potential tenants based on these retention floors as a revenue source. In addition, the savings on maintenance costs resulting from underground conservation of energy and the maintenance and operational costs to cover the increases in drain water and ventilation requirements must also be compared. In any case, it is believed that the utilization of underground space is certain to become a radical policy for urban rejuvenation.

Turning to the development of new cities incorporating both underground and surface space, a major challenge will be the construction of buildings that feature effective utilization of Level 4 space, as well as demonstration of the potential functions of such space, while not neglecting efforts to take into consideration amenities and safety.

In attempting to understand the actual state of human psychology and physiology in deep underground spaces, it will be desirable to create an environment that features long-term amenities from the viewpoint of human engineering. In order to do so, it will be necessary to eliminate the sense of enclosure associated with underground space and to heighten the sense of integration. It will be necessary to create favorable operational efficiency in such locations so that the underground environment is not inferior to environments on the surface (see Section 3.3).

Designs that are in harmony with the nature and buildings, and "*geotecture*"—a form of architecture that satisfies the architectural requirements cited in the above paragraphs—are gradually emerging as concepts that can respond to the aforementioned demands. For example, libraries, educational facilities, offices, manufacturing and storage facilities, visitor information centers, and special facilities (churches, convention centers, correctional facilities, etc.) have been placed underground successfully. From these examples, items that are particularly noteworthy are the introduction of sloped entryways, making active use of the geographical features of the land, and devices for introducing natural light. Coupled with overall measures for conserving energy, these facilities have also been appreciated for their ability to create a human-oriented environment.

4.4 TECHNOLOGY FOR UNDERGROUND DEVELOPMENT

The various technologies involved in underground utilization may be divided into three major areas:

1. Survey and design
2. Construction and execution
3. Operations and control technology

Each of these technical areas is represented in the underground development tree diagram shown in Figure 4.2. Singh and Goel [9] have discussed systematically various facets of planning, construction, monitoring, and management.

Survey and design technology is concerned primarily with the drawing up of underground maps. This area of technology can be further divided into planning technology, such as underground space data systems technology; survey technology, such as physical investigations; design technology, such as that involved in predicting the behavior of groundwater; and measurement technology, such as that related to the measurement of ground pressure.

Underground geological hazards may be predicted by Figure 4.3, depending on rock mass number N (rock mass quality Q with SRF = 1) and $HB^{0.1}$. Here H is the overburden in meters and B is the width of cavern in meters. There may be rock burst if normalized overburden $HB^{0.1}$ is more than 1000 meters and $J_r/J_a > 0.5$ and rock mass number N > 1.0. Rock mass may be in squeezing (all round failure) condition where $H >> (275N^{0.1}).B^{-0.1}$ meters and $J_r/J_a < 0.5$ (J_r and J_a are the parameters of the Q system), as otherwise there may be nonsqueezing ground (competent) condition.

It is important to know in advance, if possible, the location of rock burst or squeezing conditions, as the strategy of the support system is different in the two types of conditions. Kumar [11] could fortunately classify mode of failures according to values of joint roughness number (J_r) and joint alteration number (J_a), as shown in Figure 4.4. It is observed that a mild rock burst occurred only

---

Technology for Underground Development Planning—Survey and Design
Technology for underground planning:
- Underground information system
- Quantitative analysis of underground characteristics
- Environmental assessment

Technology for underground surveying:
- Literature search, remote sensing and surface scanning
- Physical exploration, boring and in situ test
- Investigating underground water, structure, and discontinuous bedrock

Technology for underground design:
- Bedrock modeling and use of in situ test data, and science of bedrock transformation mechanism
- Science of bedrock destruction mechanism, and technology for forecasting bedrock behavior by simulation (Fig. 4.3)
- Optimum space selection and underground water behavior
- Forecasting impact on nearby structures and reinforcing hollowed bedrock
- Analyzing discontinuous bedrock, earthquake-proofing, and geomechanics for random factors

Technology for measuring in underground space:
- Measuring ground pressure

Technology for Underground Construction
Technology for constructing foundation:
- Constructing foundation and improving bedrock

Technology for creating underground space:
- Construction, drilling, waste soil disposal, and treating the area near drilling
- Maintenance, coating, draining, and management
- Maintaining good working conditions

Technology for underground construction:
- Coping with earthquakes and other disasters
- Environmental conservation

Technology for Management
Technology for disaster prevention and escape:
- Risk management (see Chapter 15)

Technology for environmental control:
- Technology for air conditioning, ventilation, lighting, water, etc.

Technology for monitoring water, air, environment, and human behavior
Technology for maintenance of underground structures

FIGURE 4.2 Various technologies involved in underground utilization.

where J_r/J_a exceeds 0.5 and $N > 1.0$ [10]. This observation confirmed the study of [12]. If J_r/J_a is significantly less than 0.5, a squeezing phenomenon was encountered in many tunnels in the Himalaya. As a result, squeezing or rock burst is unlikely to occur where $\sigma_\theta < 0.6q'_{cmass}$ and $J_r/J_a > 0.5$ (q'_{cmass} = biaxial strength of rock mass and σ_θ = maximum tangential stress at the tunnel periphery). Figures 4.3 and 4.4 should be combined to predict whether failure will be by a slow squeezing process or a sudden rock burst.

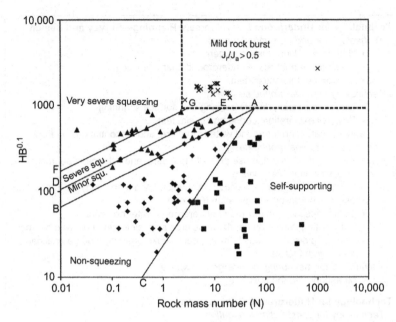

FIGURE 4.3   Plot between rock mass number N and $HB^{0.1}$ for predicting ground conditions [10].

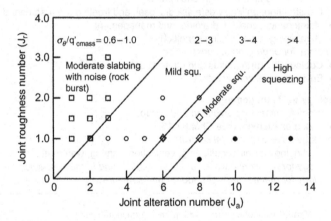

FIGURE 4.4   Prediction of ground condition [11].

The following are rules of orientation, shape, and size for the stability of (isolated) caverns and tunnels wherever feasible.

- The long axis of a cavern should be nearly perpendicular to the strike of continuous joints.
- The direction of excavation (drive) should preferably be toward the dip direction of continuous joints for stability of the face of excavation.

- The long axis of a cavern should be nearly parallel to the in situ major principal stress on a horizontal plane, which will prestress rock wedges inside the walls, if any. A set of underground openings should be planned such that the flow of stresses is smooth, minimizing stress concentrations and tensile stresses and slip zones. Openings in a highly anisotropic stress field may experience rock bursts.
- The roof of a cavern should be arched with a suitable angle (θ) with vertical (sin $\theta = 1.3/B^{0.16}$; where B is cavern span in meters). The arch will reduce the zone of tensile stresses within the roof.
- The pillar width between adjacent caverns or tunnels in nonsqueezing ground should be more than the maximum span of either cavern or maximum height of both caverns (or tunnels) from bottom to its roof, whichever is higher. It may ensure stability of the pillar and reduce seepage from one cavern to the other cavern or tunnel. However, the pillar width in the squeezing ground should be at least five times the tunnel span.
- The maximum height or span of a cavern should not exceed the upper limiting line in Figure 4.6 for a given rock mass quality, Q. The cavern needs to be located in nonsqueezing ground (overburden height H $\ll$ $275N^{0.1} \cdot B^{-0.1}$ meters). The measured modulus of deformation should be more than 2 GPa to avoid buckling failure of reinforced rock wall column. The minimum rock cover over tunnels, its portals, and over caverns should be equal to B in sound rock and 2 to 3 B in weak rocks to reduce groundwater seepage, subsidence, and vibrations of ground due to blasting.
- The intrathrust zone must be avoided in tectonically active mountains for safety. Alternatively, we can drive tunnels in the dip direction of faults and certainly not along the strike of faults. The tunnel lining through active fault zones should be segmental lining to absorb fault slip during earthquakes up to a distance equal to the tunnel width on both sides of the tunnel axis [9].
- The gradient of long rail lines should not exceed 1 in 40 in tunnels and bridges in hilly areas. Tunnel portals should be stable.
- The underground infrastructure should be planned so that risk during construction is not high; risk should be normal.

High skyscrapers may be built in the near future. Their raft or pile foundations may exert high pressures on both the roof and the walls of the supports of underground openings in soil and rock masses. The software 3DEC may give insight into interaction of foundations with openings, together with the differential settlement of their foundations.

Construction and execution technology deals primarily with improving excavation technology, such as vertical shaft and horizontal shaft construction techniques. This category may be subdivided into ground and foundation construction technology, underground space creation technology, and technology involved in the construction of underground facilities (e.g., environmental measures technology). In recent years, construction methods and technology such as shield tunneling, tunnel boring machines, and the New Austrian Tunneling Method have come into focus as fully realized tunneling construction techniques. Much underground construction is being carried out by adapting differing construction methods and numerous technologies (see Chapter 8).

One of the biggest breakthroughs in underground work has involved increasing the level of expertise in excavation technology for difficult soils. Typical technologies such as slurry walls, shield tunneling, and underpinning methods depend not only upon experience, but also on modern computer and sensor instrumentation systems. Without the development of semiconductors, breakthroughs in underground excavation would not have been possible. Observation of underground structures and the soil has become possible using highly developed sensor instrumentation systems. Analytical evaluation carried out using computer analysis permits the behavior of the soil and underground structures to be adequately monitored as construction proceeds. This breakthrough has played a major role in developing a future for deep underground construction.

Operations and control technology centers chiefly on the areas of environmental control and disaster prevention. This area of technology may be divided broadly into disaster prevention and evacuation technology for underground space, environmental maintenance technology for underground residential space, underground space monitoring technology (such as that relating to the monitoring of water and air permeability), and maintenance technology for underground structures (e.g., the repair and restructuring of underground space).

Careful studies and application of those studies will be necessary in order to apply the technology that has been used to date in creating underground space at depths from roughly 40 to 100 m.

The need is to develop legal guidelines for the public use of underground space. Governments of various countries will have to set up a committee under a ministry to plan and begin research work with private firms involved in underground construction.

4.5 CONCEPTUAL DESIGNS

A high degree of land utilization is essential in large cities. Therefore, the use of deep underground space is envisaged. In consideration of excavation techniques and size needs, dome-shaped spaces 50 m in diameter, 30 m high, and more than 50 m below the ground surface are planned. Figure 4.5 shows the concept of underground space as envisaged by Ministry of International Trade and Industry (MITI), Japan.

Conceptual studies for underground cities have been conducted by a number of firms in Japan. Most of the studies have focused on new and large-scale underground use in urban areas. The Alice City Network Plan was drafted by the Taisei Corporation of Tokyo [13].

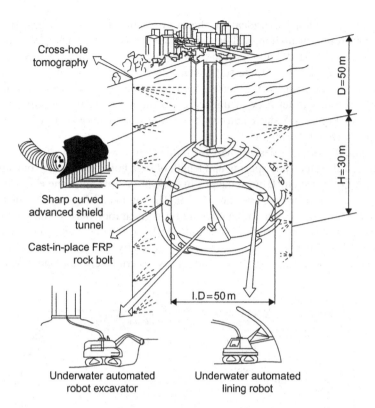

FIGURE 4.5 Concept of underground space proposed by Japan's Ministry of International Trade and Industry [13].

Named after a heroine of Lewis Carroll's *Alice's Adventures in Wonderland*, the Alice City Network Plan will result in more open space becoming available aboveground as some of the office, commercial, entertainment, information, and distribution facilities are moved underground. Alice City will be a 24-hour city in which workplaces are located near residences. Alice City will consist of the Alice infrastructure space, Alice office space, and Alice town space [13].

The Alice town space will be an area where people can enjoy themselves by going to the theater, shopping, and using the sports facilities. The Alice office space will house business facilities, shopping malls, or even hotels, connected to the surface by express elevators or an extension of the underground railway system to the bottom level. Solar dome or atrium space will eliminate the feeling of claustrophobia often associated with underground space.

The Alice infrastructure space will contain power generation, regional heating, waste recycling, and sewage treatment facilities. This underground space may be either sphere or cylinder shaped.

Underground space is ideal for the infrastructure from the standpoints of isolation, sound insulation, and earthquake resistance. It will also play a large part in preserving the aboveground environment. The Alice City Network Plan calls for Alice City to be self-contained, handling all the necessary infra-structure, office, and town functions. However, inadequacies can be supple-mented through a network of underground tunnels. The two figures at the bottom of Figure 4.5 show automatic machines for underground excavation and lining/shotcrete support.

Figure 4.6 offers a design chart for finding out the thickness of steel fiber-reinforced shotcrete, and spacing and length of rock bolts, for a given rock mass quality (Q) and span (B) of a cavern in rock in meters. Here the excavation support ratio may be taken as equal to 1.0. The top line in Figure 4.6 indicates the upper limit of the span of an opening for a given rock mass quality, Q. For example, the length of rock bolts would be 11 m for the span of caverns of 50 m in Alice City where Q > 2. Figure 4.6 is generally used to prepare a detailed project report with approximate cost estimates. Alternatively, software TM may

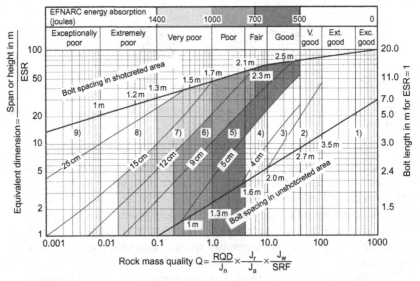

$$\text{Rock mass quality } Q = \frac{RQD}{J_n} \times \frac{J_r}{J_a} \times \frac{J_w}{SRF}$$

Reinforcement categories

1) Unsupported
2) Spot bolting, sb
3) Systematic bolting, B
4) Systematic bolting and unreinforced shotcrete, 4 to 10 cm, B (+S)
5) Fiber reinforced shotcrete and bolting, 5 to 9 cm, S(fr) + B

6) Fiber reinforced shotcrete and bolting, 9 to 12 cm, S(fr) + B
7) Fiber reinforced shotcrete and bolting, 12 to 15 cm, S(fr) + B
8) Fiber reinforced shotcrete > 15 cm, reinforced ribs of shotcrete and bolting, S(fr), RRS + B
9) Cast concrete lining, CCA

FIGURE 4.6 Chart for the design of support including the required energy absorption capacity of steel fiber-reinforced shotcrete suggested by [14,15].

be used to design the support system for caverns and tunnels with or without wide shear zones and seepage pressures [12].

While planning underground space use, it is necessary to take care of the general problems of human beings. Problems can be tackled by considering the following points [16] (see Chapter 3).

- *To deal with phobias about being underground:* Create entrances and exits that have a natural connection with the surface; supply natural lighting; provide an airy, light interior design.
- *To deal with an unhealthy environment:* Provide proper temperature and humidity control, and ventilation. The latter relates to air conditioning, not only in terms of maintaining a certain temperature, but also in terms of being able to adjust the temperature in different rooms and to increase ventilation rates.
- *To deal with the lack of orientation in the underground:* Create designs that impress themselves upon people's senses of sight and hearing. For example, the use of dissimilar columns (i.e., carved, designed, and marked in different ways) makes it possible for people to confirm the direction in which they are heading. In a fire emergency, people can exit safely through the smoke by touching these columns.
- *To deal with disaster prevention and safety:* Use the space in a manner that avoids risks to the ground above. One step in this direction is to develop software systems as well as hardware systems (such as high-technology machine systems).
- *To deal with the lack of visual stimulation:* Provide a variety of activities in the buildings and/or promote creative visual designs that will satisfy the need for visual variety.
- *To deal with the lack of natural light:* Research solar day-lighting systems that provide rooms with natural light collected through a system of mirrors and fiber scopes; incorporate atriums and sunken courtyards into building designs.

Light having the same wavelength as that of natural light has been produced for use in conjunction with office, home, and atrium systems to produce the most comfortable environment possible [8].

Information systems will include large-screen, high-definition televisions carrying reports of surface activities and digital displays of surface temperature and humidity. Information will be available on a 24-hour basis to decrease the feelings of isolation that might occur from living or working underground.

To mitigate feelings of confinement, high-ceilinged rooms will be developed in most areas, along with pseudo-windows that will create various outdoor environments, such as mountain or lake areas, through the use of sensory and olfactory devices. Recreation atrium areas will use trees and fountains to create park-like atmospheres. Ventilation and air-conditioning systems will be of superior quality.

4.6 COST CONSIDERATIONS

Although underground construction is more expensive than surface construction, the cost of land decreases with depth. Figure 4.7 shows the decrease in land costs with depth based on a subway construction in Tokyo. Because of this decrease, additional construction costs will be balanced by comparatively small land costs, as shown in Figure 4.8.

Private building owners usually do not engage in deep underground construction. Consequently, deeper zones are expected to be used for urban infrastructure, even though the matter of monetary compensation for overlying landowners is an issue that remains to be solved. Recently, attempts have been made to establish guidelines limiting private land ownership in the deeper underground. It is hoped that new guidelines will permit development to proceed without public developers having to purchase deep underground land.

Variation of cost of tunneling with rock mass quality Q is discussed in Section 8.8.3.

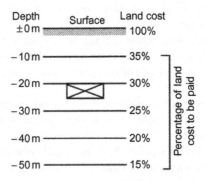

FIGURE 4.7   Decrease in land costs with depth based on subway construction in Tokyo [13].

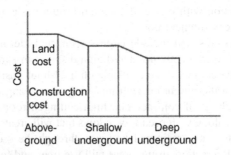

FIGURE 4.8   Comparative land and construction costs for aboveground, shallow underground, and deep underground structures [13].

4.7 PLANNING OF UNDERGROUND SPACE

The first problem relates to the preparation of a long-term master plan and the establishment of a flowchart for the utilization of underground space. In addition to presenting the inverted dome form as the ideal urban form to be pursued, it is critical to establish a master program aimed at the realization of such a form. Second, it will be necessary to conduct assessments of effective utilization, incurred costs, environmental factors, safety, and amenities for underground space under both normal and emergency conditions. In addition, a comprehensive evaluation of underground use, taking into consideration both present and long-term transitions, will be required. In order to achieve this, it would be desirable to create life-size test buildings and then to establish actual evaluation procedures, based on human engineering standards, for underground space Levels 1–4 (Fig. 4.1).

The third topic requiring attention is the need to pursue systematic technology that reorganizes the potential utilization of shallow layers of underground space, thereby creating a rational sequence that includes deep underground space as a technology that can actually accommodate such utilization. An example is the use of new construction methods for ducts and underground parking lots, etc. In order to attain this goal, it will be necessary to abandon those items that are not necessary, to repair those that are necessary, and to move new structures underground while devising methods that will serve to give such structures flexibility and allow multifunctional utilization.

Fourth, in order to contribute to the rejuvenation of urban and rural areas, systems that ensure proper operation and control of underground space will be required. In order to accomplish this, it will be necessary to improve systems aimed at securing the safety of occupants. Implementing a centralized control through the monitoring of each level on a real-time basis, using video equipment, will be desirable as an effective means of management and control of underground facilities.

Finally, studies relating to improvements of operations combining geotecture and redevelopment, as well as the integration of these concepts into potential sites (such as waterfront sites), will be required. Additional topics, such as the conformity between individual uses and the overall plan, and methods of developing underground space into rational, multidimensional cities should also be examined in the future.

Efficient strategies in city planning may focus on utilization of the underground space. It is evident that the number of inhabitants is not the only important factor to be addressed in estimating town space needs. Another factor is the amount of space available for future town development.

The shortcomings of local policies and the need for more far-sighted planning have been stated by [17]: "Our world, with its limited resources is beginning to realize that local planning ought not to be carried out in a vacuum. Place things below the surface, and put man on the top."

Two different types of plans may be prepared for underground space: (a) a project plan and (b) a city plan. The plans differ in terms of the validity of planning stipulations and notations.

The project plan is prepared at the point when it is certain that the underground project will go ahead. The city plan shall be drawn up at the same time as the project plan. At this stage the plan may go into quite considerable detail in terms of notations and stipulations. City planning is a dynamic program that is basically a trial-and-error approach or design-while-you-go approach.

Project plans need to be backed up with more general planning to make sure that the various aspects related to land use in the vicinity of the project are given proper attention.

In Helsinki, the details of upcoming projects are nowadays known fairly well in advance, and for the city of Helsinki (in its capacity as a significant land owner) it is fairly easy to coordinate city plans with project implementation in the case of both surface and underground plans. This is known as project planning, that is, the city plan is tailored according to the upcoming project. However, there are also cities and municipalities where the situation is very different and where projects do not go into the implementation stage until several years after the plans have been completed. For such cases, too, it is necessary to make the appropriate reservations in city plans for underground space. The city plan can then be revised according to the requirements of each project if the stipulations are no longer in force or if they deviate from the project plan.

A fairly loose (liberal) city plan is usually broader than a project plan and is prepared at an earlier stage in the planning process than a project plan. This kind of city plan is required when there is no general plan for land use in the area concerned and when, with underground construction in mind, existing space and future needs will require further land reservation.

A loose city plan shall be less specific in its notations and stipulations than a project plan so that when the project moves on to its implementation stage, it will not be necessary to revise the city plan (which is a time-consuming business). A loose flexible city plan affords greater responsibility for implementation to local building inspectors (e.g., in the planning of portals and ventilation ducts).

At the stage of project planning, bedrock inventories are conducted to optimize land use and to minimize costs. If a detailed bedrock inventory has not been carried out for a loose city plan, project feasibility in terms of rock engineering will have to be assessed on the basis of maps. Therefore, the city plan, which is ratified, must allow for sufficient flexibility.

Separate city plans may be prepared for different levels in the process of city planning. Its purpose is to provide a clearer picture of the planned use of underground space at different depths.

REFERENCES

[1] Barles S, Jardel S. Underground urbanism: comparative study [l'urbanisme souterrain: étude comparée exploratoire]. Paris: Université de Paris 8; 2005.

[2] Besner J. Genesis of Montreal's indoor city [génèse de la ville intérieure de Montréal]. Proc of 7th International Conference of ACUUS, Montreal; 1997.

[3] Sijpkes P, Brown D. Montreal's indoor city: 35 years of development. In: Conference of the associated research centers for urban underground space, Montreal; 1997.

[4] Seiki T, Nishi J, et al. Classification of underground space and its design procedure in Japan. In Conference of the associated research centers for urban underground space, Montreal; 1997.

[5] Takaksaki H, Chikahisa H, Yuasa Y. Planning and mapping of subsurface space in Japan. Tunnel Undergr Space Technol 2000;15(3):287–301.

[6] Ronka K, Ritola J, Rauhala K. Underground space in land-use planning. Tunnel Undergr Space Technol 1998;13(1):39–49.

[7] Vahaaho IT, Korpi J, Anttikoski U. Use of underground space and geoinformation in Helsinki, ITA Open Session, Singapore; 2004.

[8] Watanabe Y. Deep underground space: the new frontier. Tunnel Undergr Space Technol 1990;5(1/2):9–12.

[9] Singh B, Goel RK. Tunnelling in weak rocks. In Hudson JA, series editor. Geoengineering series. UK: Elsevier Ltd.; 2006. p. 489.

[10] Singh B, Goel RK. Engineering rock mass classification: tunneling, foundations and landslides. USA: BH Elsevier; 2011. p. 365.

[11] Kumar N. Rock mass characterization and evaluation of supports for tunnels in Himalaya. Ph.D. thesis, W.R.D.M., IIT Roorkee, India; 2002. p. 295.

[12] Singh B, Goel RK. Software for engineering control of landslide and tunnelling hazards. Lisse: A.A. Balkema Publishers (Swets & Zeitlinger); 2002. p. 344.

[13] Hanamura T. Japan's new frontier strategy: underground space development. Tunnel Undergr Space Technol 1990;5(1/2):13–21.

[14] Grimstad E, Barton N. Updating of the Q-system for NMT. In Kompen R, Opsahl O, and Berg K, editors. International symposium on sprayed concrete, modern use of wet mix sprayed concrete for underground support. Oslo: Norwegian Concrete Association; 1993.

[15] Papworth F. Design guidelines for the use of fibre-reinforced shotcrete in ground support, shotcrete. Web link: www.shotcrete.org/pdf_files/0402Papworth.pdf; 2002. p. 16–21. Accessed in 2008.

[16] Carmody JC, Sterling RL. Design strategies to alleviate negative psychological and physiological effects in underground space. Tunnel Undergr Space Technol 1987;5(1/2):59–67.

[17] Barker MB, Jansson B. Cites of the future and planning for subsurface utilization. Undergr Space 1982;7:82–85.

Chapter 5

Underground Storage of Food Items

Although the world is full of suffering, it is full also of the overcoming of it.

Helen Keller

5.1 GENERAL

A strong link is emerging between the development of new materials and ancient agricultural practices. Underground food storage is one offspring of this link. This is being forced by the costs of capital investments in new structures, recurring energy needs, and repeated applications of pest management chemicals associated with aboveground storage. The development of new materials for underground structural support and waterproof, airtight sealing are making "old" alternatives possible.

Underground space has been used by human beings (and other mammals) for millennia as a place to store food safely. Choosing the proper soil and drainage conditions and preparing the pits well in locations throughout the world have served farmers in preventing postharvest loss. People in ancient cultures understood that cool, stable temperatures of a sealed environment provided important advantages. Reduced losses from fire and reduced pilferage by insects, rodents, birds, and humans were other attributes associated with underground storage structures.

In China (600–900 A.D.), the national grain reserves were stored in a large array of underground pits beneath the capital city of Lou Yang (in present-day Henan Province, China). The low water table and loess soil of the region were ideally suited to this type of construction. When in use, the facility contained 1.25 million T of rice and millet. Documented subterranean storage in the Middle East and northern Africa has an even longer history. Typical on-farm storage in Mediterranean areas involved flask-shaped pits that were often carved into rock or lined with locally available waterproofing materials. In other areas, such as sub-Saharan Africa (Nigeria, Somalia, and Ethiopia) and India, farmers also have traditionally used underground pits for grain storage. North American farmers have used underground space, in the form of root cellars, primarily for the storage of tubers rather than grain [1].

In the 20th century, agricultural research has markedly improved aboveground metal bins, warehouses, hangers, and godowns. Through efficient extension services at land-grant institutions in the United States and with technical assistance from many countries, these structures have been promoted throughout the developed and developing world for the safe storage of bulk grain and legumes.

A new era for storage of food grains is now emerging because of technological advances in the use of underground space. The forerunners of this new approach may be found in South America. During World War II, the threat of losing foreign markets for their accumulated grain stocks forced the Argentinean government to allow experimentation with underground, large-scale, long-term storage. A total of 1474 underground silos, each with a 600-T capacity, stored the farmers' wheat throughout the war and kept it in excellent condition until foreign markets reopened.

In Brazil, farmers are now having success with small-scale polyethylene-lined pits for corn and bean storage. After successful trials with corn in underground storage in Minnesota, the technique was introduced for bean storage in Rwanda through a project sponsored by the U.S. Agency for International Development. In successful experiments in a wet clay soil, utilizing locally produced polyethylene, beans maintained their fresh appearance and no insect development occurred. The corn in Minnesota, even at 16% moisture content, was similarly free of insect and fungal development for 3 years [1].

In China, new spherical-shaped structures sealed with bitumen have been built near the ancient underground storage site in Henan Province. Unique loading and unloading features make this large-scale facility quite efficient. In the United States, limestone caverns under Kansas City are being used for storing millions of metric tons of grain as well as for storing processed foods.

Food items have been classified into two categories from the point of view of the basic requirement of temperature and humidity/moisture in the stored environment. These are (i) food grains, which are generally stored in normal temperature conditions, and (ii) fruits and other items that require cold/refrigerated conditions to store.

Section 1 discusses the underground storage for food grains, whereas Section 2 covers the storage of fruits and other perishable food items in refrigerated conditions.

SECTION 1

5.2 PROBLEMS ASSOCIATED WITH UNDERGROUND FOOD STORAGE

Clearly there are potential cost advantages in using the earth for grain storage, that is, using the earth for structural support, to assist in achieving airtight conditions, to prevent pilferage, and to maintain a constant temperature and equilibrium relative humidity. However, the approaching new era in underground food storage will require us to solve several major problems, including:

• The lack of accurate, economically feasible remote-monitoring systems for determining critical physical and biological parameters

- The lack of understanding of fungal species that thrive in a low O_2 and high CO_2 atmosphere of such an environment and the influence they have on weight loss and toxicity
- Psychological barriers to having the products of harvest in unseen storage containers

A solution of these major constraints to the development of underground food storage for the prevention of postharvest loss will require an interdisciplinary effort. Material engineers will need to collaborate with biologists, agricultural engineers, and social anthropologists. This research arose out of a spirit of cooperation and professionalism. The same spirit of cooperation and professionalism, applied to the specific use of underground space in the area of food storage, may result in important benefits to the international community [1].

5.3 SITE SELECTION

Underground facilities for grain storage should be located where transportation costs and transportation times can be minimized. A facility at a major port where imported grain arrives is an ideal location, provided that the rock formations near the port are suitable for construction of such a facility.

Another good choice of sites for large-scale underground or rock cavern storage of food would be facilities adjacent to the centers of population where most of the consumption will occur. Within a metropolitan area, the best location would be adjacent to a processing facility, such as a flour mill or rice polishing unit. Some large metropolitan centers, such as Kansas City and Minneapolis/ St. Paul, overlie or are adjacent to a network of extensive limestone caverns. Other large metropolitan areas, such as Seoul, Korea, and Oslo, Norway, have formations of granite and gneiss that are particularly suitable for the construction of such underground storage structures. Most major metropolitan areas, however, are built in estuary areas where the rock base is soft, unconsolidated, and water-saturated. Another consideration involves making use of the slope of hills to conserve energy. The facility could be designed so that grain arrives at the top of the facility and bins are filled by gravity. Bins can then be unloaded at the base of the hill, also by gravity. Groundwater depth is another consideration. The groundwater table should lie well below the structure being considered. Large cities are often built on old river deltas or sides of rivers, where the water table is high. Therefore, major metropolitan areas may not be suitable for underground grain storage [2].

5.4 CONSTRUCTION AND DESIGN

5.4.1 Depth

Sealed (airtight), underground storage at a sufficient depth below the surface prevents the major causes of loss in grain by

- Maintaining a uniform low temperature
- Establishing and maintaining a low oxygen atmosphere
- Minimizing accessibility to the grain (Fig. 5.1)

Stored grain losses are minimized by these key biological factors without the use of residual or fumigant pesticides. These three factors, which make the underground environment a wise choice in selecting grain storage structure designs, are discussed here.

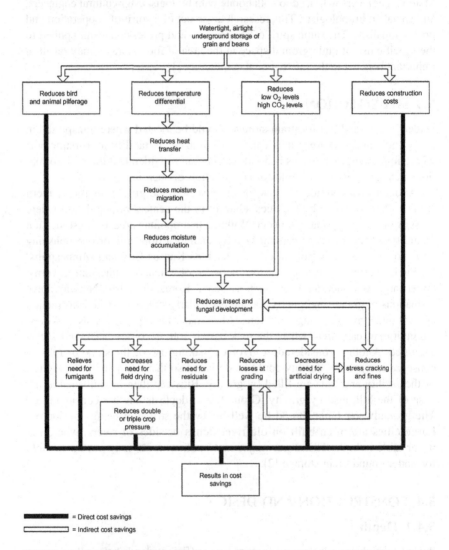

FIGURE 5.1   Flowchart of direct and indirect benefits and cost savings associated with storage of grains, beans, and other commodities in underground, airtight, and watertight structures [2].

Uniform Low Temperature

At a certain depth in the earth or in a rock formation, the soil or rock temperature ceases to fluctuate with ambient aboveground temperatures. The particular depth at which this phenomenon occurs depends on many factors, including the geological characteristics of the site. The level at which the storage facility is placed depends on this geologically determined temperature pattern and the economics of the excavation process: the correct depth is unique to the specific location, but generally it may be taken as 8 to 10 m below the ground surface.

Except in areas of geothermal activity, this constant soil or rock temperature throughout the world is best estimated as slightly above the mean (average) annual air temperature. The constant value of annual air temperature depends on several other conditions. For example, in Minneapolis, Minnesota, the annual average air temperature is about 7.5°C (45°F). Snow cover and the warming effect of the metropolitan area raise local ground temperatures to about 13°C (55°F). In some areas of Korea, the constant value is close to 13°C (55°F) [2].

Accumulation of moisture in one or more locations in stored grain encourages the growth of both fungal and insect populations. A constant and uniform temperature of the grain and the surrounding environment throughout the storage period will prevent temperature gradients, which, in turn, will prevent moisture migration and accumulation. A low temperature (e.g., 13°C) will prevent insect reproduction: that is, this temperature will not kill insect and mite populations, but they will not increase in population.

Low Oxygen Content in the Space between Grain Kernels

At levels of oxygen (O_2) much lower than ambient air (20%), particularly at about 2 to 3% O_2, insects and mites will not survive. Some insect species (e.g., grain borers, the angoumois grain moth, weevils, and bruchids) constitute a "hidden insect infestation" because most of their life cycle occurs while the insects are locked inside the grain kernel or bean seed. In a hidden infestation, there are no easily observable external signs that the grain is infested. Most of the actual damage to the grain or bean kernel is done by insects in the immature stages, specifically all stages of the larva. The larva has a worm-like shape with smaller legs.

Adults of borers and weevils can also cause damage to the grain. Although mites live outside the grain, they are small (0.1 to 0.5 mm) and easy to overlook in grain inspection at an ocean port or anywhere else in the grain marketing chain. Grain bulks may be brought in from the field or placed into ship holds for trans-oceanic shipment in seemingly excellent condition, while there is actually a hidden insect or mite infestation. If conditions are favorable for these insects or mites, the grain may be seriously damaged by the time it reaches its destination. Under ideal conditions, 4 weeks is the shortest time for a complete life cycle of stored grain insects and mites. Therefore, if the grain has a visible insect problem or if a hidden insect infestation is suspected, it is advisable to place the grain immediately into underground storage, which has a fumigative effect (associated with the decreased O_2).

Nanobiotechnology in the near future may revolutionize the genetically modified, vast variety of food grains, big fruits, and vegetables, which may protect themselves from all insects during growth as well as during underground storage. Genetically modified agricultural products may be very nutritious and durable in the near future.

Reduced Accessibility

Grain stored in a sealed underground structure will be inaccessible to birds, and it may be made inaccessible to humans. If grain storage facilities are constructed in rock formations or within properly designed structures in the soil, they will also be inaccessible to burrowing mammals. If grain or another stored, dry commodity is difficult for humans to access, it is likely that the commodity will be mobilized only when necessary. Easily accessible grain or other dry commodities are ready sources of cash. For individual producers, if grain is easily accessible, it is often sold first, before other alternatives to a low cash flow are utilized. Similarly, because of its cash value, accessible grain is stolen and converted to cash easily.

Grain emits odors. In aboveground storages, these odors are detected easily by stored grain insects. These insects are excellent fliers and their olfactory senses are keen. A sealed underground facility prevents such "advertising" and subsequent insect entry into the structure. Not only is an underground, sealed storage facility inaccessible to living organisms that consume the grain, but it also can be made inaccessible to conditions destructive to the grain, such as fire, nuclear fallout, rain, groundwater, and ground to cloud lightning. Large-scale aboveground storages, particularly those with a grain elevator system, are prone to explosions [2]. Explosions and fires in grain bins and grain elevators are caused by the combustion of grain dust suspended in air.

When three conditions—uniform low temperature, low oxygen, and reduced accessibility—are met, underground grain storage will provide additional benefits, including:

- Improvement of environmental and consumer safety
- Economic advantages
- Wise use of natural resources

Environmental and consumer safety are improved by prevention of the growth of insects, mites, and fungi without the use of residual or fumigant pesticides. Economic advantages include a reduction in both the costs of construction of the storage facilities [3,4] (Fig. 5.1) and the long-term preservation of grain quality. The ability of the storage system to preserve the quality of the grain over the long term will allow the government or private sector to buy when grain prices are low and store the grain until grain prices are high.

5.4.2 Unit Size

Because the grain storage structures must be sealed and retain a uniform temperature, multiple entry is not advisable. If a portion of grain in storage is needed for consumption or for transportation to another location for processing, the seal must be broken. Once the seal is broken, the entire contents of that unit should be used, as the safe conditions for underground grain storage will have to be reestablished for the remaining stored grain in the structure.

If the grain is stored in sufficiently small sealed structures, it is likely that the amount of grain that needs to be mobilized at one time will not be smaller than the size of one entire storage unit. Then, each time the seal is broken in a storage unit, the entire bulk can be utilized. The size of the structure should be optimized using linear programming or some other optimizing technique based on expected usage times and quantities. For underground storages, designers should consider multiple, smaller units than would be appropriate for aboveground storages used for the same purpose.

5.4.3 Lining

If a thin plastic liner is selected as the construction material, a subterranean rodent barrier must be added to the design. However, if storage units are built in friable calcareous limestone similar to that of North Africa or in rock formations such as the granite or granite gneiss found in Korea and Norway, rodent barriers will not be necessary. It is essential that any liner serves as an absolute barrier to ground-water seepage. To prevent seepage from occurring, the strength of the liner, the amount of care taken in handling the liner when placing it in the bin, and the nature of the bin wall that it lines; all need to be considered.

5.4.4 Loading and Unloading Equipment

The most important consideration in designing the loading and unloading process for underground structures is accessibility. Cliff and mountain locations may be easy to excavate and may permit bins to be filled and emptied by gravity. However, cliff and mountain sites may be a long distance from production, port facilities, or areas of stored product use.

Another important consideration for any large-scale grain storage facility is the distance that the grain must drop in the loading process. Impact force can cause breakage; broken kernels go out of condition faster than sound kernels. Moreover, moisture, insects, mites, and/or fungi in out-of-condition areas can diffuse into and contaminate adjacent sound grain.

5.4.5 Equipment to Equilibrate Grain Temperature during Loading and Unloading

It is essential that the grain that will be placed into the underground storage is at the same temperature as the rock or earth surrounding the structure. Earth temperature varies from region to region depending on the annual average temperature. There are several ways to achieve this requirement, but each requires planning prior to the time of loading. The least costly—although perhaps inconvenient—procedure is to load the structure when the grain is at the temperature of the earth. If the grain is coming directly off of a transoceanic freighter, this temperature is difficult to predict. If the grain has been held after harvest in local storage or port storage, the best time of year to obtain grain is at about 13°C, which is in the fall and spring. Grain is an excellent insulator and, therefore, will warm slowly and cool slowly. The time at which the majority of the grain mass is at 13°C must be monitored closely.

A second means of achieving grain temperature is to use an aboveground chilling device. Refrigeration units have been developed and are in large-scale experimental use in the United States (Indiana and Michigan), Indonesia, Australia, and Germany [5].

Both heating and cooling units designed for aboveground grain management can be used to equilibrate grain during the loading and unloading of underground structures. Low-temperature grain dryers can be used to increase the grain temperature in small increments. The grain-chilling apparatus forces ambient air over a bank of refrigeration coils, and then the grain, in order to decrease the grain temperature [5]. Because dry grain will easily absorb moisture from the cool, wet air, the air is reheated a few degrees to reduce the relative humidity to a range of 60 to 75%. More than 6.7 MT of grain on four continents are chilled by these methods annually [2].

5.4.6 Considerations while Loading the Storage

Grain should again be checked to confirm that it is of good quality, that is, free from observable insect infestation (insects live or dead, insect-produced odors, insect-damaged kernels) and low in moisture content. Each grain or stored oilseed or legume seed has a different equilibrium moisture content, given a specific interkernel relative humidity and grain temperature. A table of safe storage life [6] has been developed to assure safe storage at a specific grain temperature and grain moisture content.

For wheat or rice, a moisture content of 12.5% is the maximum advisable limit for 1 year of storage at 13°C with normal O_2. The storage time is indefinite (i.e., many years) if an O_2 of 2 to 3% can be maintained throughout storage. Often, grain will arrive from an overseas shipment with an insect infestation that has developed during the transoceanic journey. Sometimes the composite moisture content of the grain to be stored is slightly higher than ideal. It is possible to store this grain safely in sealed, underground facilities that prevent any further deterioration in the quality of the grain [2].

In sealed storage, further deterioration is prevented in insect-infested and/or fungal-infested grain or beans. Similarly, the harmful effects of medium-moisture conditions are arrested because the insects and fungi that would thrive are inhibited by the low O_2 atmosphere and the low grain temperatures. Of course, the quality of the grain can never be improved unless it is blended with other, higher quality grain. The temperature of the grain at the time of loading the storage facility, as well as the temperature of the interior of the storage structure, particularly the walls and floor, should also be monitored carefully.

A final assessment of the specific temperature management plan, based on the considerations prior to loading (described in the preceding section), should be made. Just prior to loading, the advisability of proceeding should be confirmed on the basis of temperature.

5.4.7 Considerations prior to Unloading of Underground Storage

The most important considerations prior to unloading grain are to estimate what the differentials in temperature of grain versus ambient temperatures will be and to develop a plan to ensure the safety of workers in newly opened bins. If the grain must be mobilized in midwinter or midsummer, it is likely that a grain warming or grain cooling plan, such as that described earlier, will need to be implemented.

Worker safety is important because of the modified atmosphere created by this type of storage. When the seal is broken, workers should refrain from entering the storage or climbing on the surface of the grain until a normal atmosphere (0.3% CO_2 and 20% O_2) is achieved.

The approach of the gas mixture to normal can be monitored by hand-held colorimetric glass tube indicators.

5.4.8 Underground Storage Bins in Argentina

In Argentina, 1540 underground silos with cement floors and walls, with a total capacity of 849,300 MT, were used to store corn, wheat grain, and wheat flour. Each silo had a capacity of 600 MT. The silos were built in nine areas of the grain-producing regions. These regions had primarily hot and moderately damp climates.

Wheat produced in Argentina was stored in 1187 of these silos (Figs. 5.2(a) and 5.2(b)). The wheat was monitored carefully for changes in temperature, moisture content, and CO_2 of the interkernel space. Data from germination, organoleptic analysis, and fat acidity and from milling, dough, and baking property analyses indicated that this wheat, totaling 688,866 MT, was preserved for 1.75 years in excellent condition (0.45% loss) [2].

Experimental designs of these bins (Fig. 5.2) included:

- Soil–cement wall and floor structures
- A grain-covering design involving two impermeable, flexible liners with a reed mat in between
- An asphalt emulsion for sealing this cover design with the soil–cement wall

Some experimental designs also included a coat of reflective paint on the very top layer.

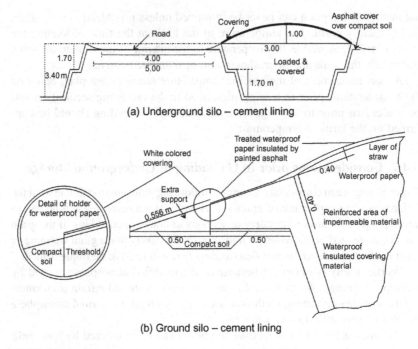

(a) Underground silo – cement lining

(b) Ground silo – cement lining

FIGURE 5.2   Underground storage bins for wheat in Argentina [2].

5.5 POTATO STORAGE

In contrast to grain storage, potatoes cannot be stored in airtight conditions, as an oxygen supply should be maintained to all regions of the potato storage bin. For potatoes intended for consumption (i.e., not seed potatoes), ideal conditions are to store the potatoes in the dark at temperatures of about 5–10°C. Low temperature results in slow sprout growth and reduces losses caused by potato respiration. In addition, pathogen metabolism is slowed at reduced temperatures so rotting is often controlled [7]. It is important to maintain relative humidity close to 90%, but condensation should be avoided. The optimum condition for storage is closely approached by underground storage methods in the absence of mechanical cooling equipment; even when such equipment is available, energy costs of operation would be substantially less with underground storage.

SECTION 2

5.6 REFRIGERATED ROCK STORES

One of the most economical uses of underground space is for refrigerated storage located in rock. Cold storage accommodates a variety of storage objects, including meat, fruits, vegetables, or any other food storage that needs to remain cold or frozen, such as packaged food or seafood that requires refrigeration to remain fresh. In addition to food, one can store other heat-sensitive storage items such as valuable artifacts, rare paintings, films, movies, and many other items that may need cold storage. Underground cold storage offers additional security measures to prevent unauthorized access to the cold storage.

Such underground cold storage in rocks has been built in Sweden, Norway, and Japan, as well as in Finland. The refrigerated store is worth locating in rock because the rock itself is an insulating material and, thus, keeps construction costs low. The space does not need to be insulated; instead, the rock surface itself or shotcreted surfaces are used. The required capacity of the refrigeration equipment is small because the rock has a constant initial temperature of approximately +7°C throughout the year in Scandinavia and other cold countries [8]. It may be noted here that the rock temperature varies from region to region depending on the average annual temperature.

In order to reduce power consumption and effectively control the temperature inside a food storage area, on ground or underground, the space should be as insulated as possible. Obviously, food storage tanks at the surface would only yield a certain degree of insulation, whereas in underground conditions, control of the environment would be achieved in a much more efficient manner, especially when the storage cavern is excavated in a rock mass having a low thermal conductivity [9,10].

Food storage could be divided into two main categories: namely chilled (temperature around 0°C) and frozen food storage (much below 0°C, say around −25°C). In the case of frozen food storage, the temperature difference between inside the cavern and the surrounding environment is high, consequently leading to a higher amount of heat transfer. Therefore, energy consumption would obviously be more critical in the case of frozen food storage.

In Scandinavia, the energy consumption for deep freezer storage is 75% and for refrigerated storage only 25% of similar surface stores. Peak energy requirements, and thus installations, are even more favorable. Deep freezer storage will need 50% and refrigerator storage only 20% capacity of similar surface stores [11].

Highly reduced insurance rates are also favoring the subsurface solution for cold stores. This is due to the fact that the rock mass surrounding the storage caverns contains a big cold reservoir. In case of a breakdown in the cooling machinery for a couple of weeks, an increase in the temperature of only 2−3°C is measured [11].

5.7 DESIGN OF UNDERGROUND COLD STORAGE

5.7.1 Shape and Size

Results of a study regarding the size of a refrigerated rock store from a thermal point of view show that a bigger size is better. However, size has often been limited from the rock mechanics point of view. This has been overcome

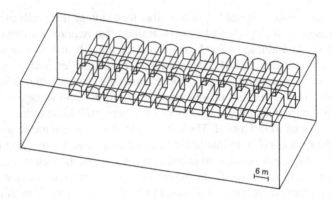

FIGURE 5.3 A three-dimensional model of a modern underground cavern [12].

by placing several tunnels side by side. The layout of a modern cold storage underground cavern is shown in Figure 5.3. It generally has a main access tunnel from which there are several cross tunnels side by side. The length of the main tunnel and the number of branch tunnels depend on the capacity of the storage facility. The height and the width of the branch tunnel can also be decided on the basis of rock quality and the space requirement.

The location of the refrigeration plant may be at a place convenient to the handling and operation points.

For a refrigerated (chilled or frozen) store in rock, insulation installed in conventional stores is replaced by the large masses of the surrounding natural rock. The existing heat accumulated in the rock masses is withdrawn during an initial period before the store is ready for use. When, however, the design temperature is reached, the rock acts as an efficient heat and moisture barrier that improves continuously over the years of operation. Due to this characteristic behavior, proper thermal design of rock stores is more important than for conventional stores.

5.7.2 Thermal Properties

Thermal properties of rock, especially thermal conductivity, may have an important influence on the performance of a cold store in rock. Thermal conductivity depends on the rock type, for example, the minerals and texture of the rock [13]. The most dominating mineral from a thermal point of view is quartz, with a thermal conductivity of approximately 8 W/mK. The lowest conductivities are found in feldspar, mica, and calcite, which are in the range of 2.0–3.0 W/mK. This means that the lowest conductivities from 2.0 to 2.5 W/mK are found in mica schist and limestone, producing favorable conditions for a freezer store. The highest values of 3.5 to 6.0 W/mK are found in granite and quartzite. Gneissic rocks are normally of intermediate quality from a thermal conductivity point of view.

5.8 COST COMPARISON

By modeling of heat transfer details around a surface, food storage tank, and underground caverns excavated in tuff and granite, it was possible to make a comparison on the basis of operational costs depending on energy loss. An underground storage cavern built in tuff is superior over other alternatives. A comparison between storage caverns in terms of construction cost was also made. Table 5.1 presents a rough estimate of the overall construction costs of three alternatives by considering current conditions in Turkey [12].

The actual cost for the Kastbrekka cold store in rock is given in Table 5.2 as per the cost level of 1978 [13]. The store has the main dimension of $W \times H \times L = 15 \times 8.5 \times 85$ m^3, which gives a volume of about 10,000 m^3. The rock is a greenstone with a thermal conductivity of 3.0 to 3.5 W/mK. This is an important case history that highlights the undersize of refrigeration plant and overcoming this problem [13].

The operating costs for a refrigerated store in rock are also lower than those of equivalent surface stores (Fig. 5.4), primarily because of their lower energy costs. The service requirements of subsurface space are also lower. Refrigerated

TABLE 5.1 Comparison of Approximate Overall Construction Costs of a Surface Food Storage Tank and Underground Storage Caverns Excavated in Granite and Tuff [12]

Type of Storage	Construction Cost (US$)
Surface food storage tank	16,000
Underground storage caverns excavated in granite	16,000
Underground storage caverns excavated in tuff	6000

TABLE 5.2 Costs for Kastbrekka Cold Store [13]

	Cost	
Cost Factor	NOK[a]	US $
Rock store excavation	3,000,000	428,600
Refrigeration machinery	450,000	64,285
Racks for goods	286,000	40,857
Additional costs, including financial costs	1,171,000	167,288
Total cost	4,907,000	701,000

[a]Norwegian krone.

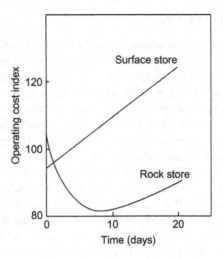

FIGURE 5.4 Relative operating costs in refrigerated stores [8] (time in days).

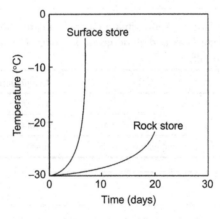

FIGURE 5.5 Rise in temperature after power failure in refrigerated stores [8] (time in days).

rock stores are safe and reliable because the ambient temperature changes very slowly in the event of a breakdown (Fig. 5.5).

A refrigerated rock store in Velkua is used for storing fish. The storage system, designed to contain all the cooling functions underground, consists of two parallel halls, each 12 m wide and 33 m long. The electricity substation and the cooling equipment are in a separate hall off the access tunnel. The equipment is designed to provide an operating temperature of −27°C. Fish are stored in truck pallets stacked on top of each other; racks are not needed. The four compressors for the cavern drive the plate freezers and the air coolers. The condensers are located aboveground, where tubes are led through a shaft. Freon is used as the refrigerating agent in the refrigerating equipment.

Construction costs for a refrigerated rock store vary, depending on the size of the store, between 800 and 1200 Fmk/m^3 for cases similar to the Velkua store; this cost includes plate freezers.

It is inevitable that use of the underground for civil purposes will increase in the near future. It is believed that with an increase in both chilled and frozen food storage in underground caverns equipped with fully automatic environmental control facilities, preservation of food could be performed in a more efficient manner, leading to a supply of high-quality products to consumers. This is underlined by the sudden increase in the cost of grains in 2008, as corn is being used for generating ethanol, which is mixed in petrol for reducing its cost; the prices of crude oil are soaring uncontrollably to US$140 and more per barrel (from US$30 in 2003).

5.9 CASE HISTORIES

5.9.1 Warehouse Caverns in Singapore

The scheme consists of five warehouse caverns and an optional additional cavern as a civil defense bomb shelter. The layout is shown in Figure 5.6. Each of the five storage caverns excavated in Bukit Timah granites in Singapore is 20 m wide, 22 m high, and 118 m long. The parallel storage caverns are

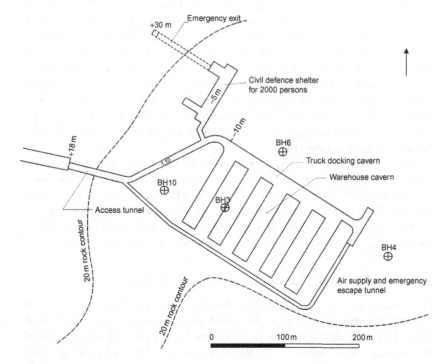

FIGURE 5.6 Layout of the proposed warehouse and shelter caverns in Singapore [14].

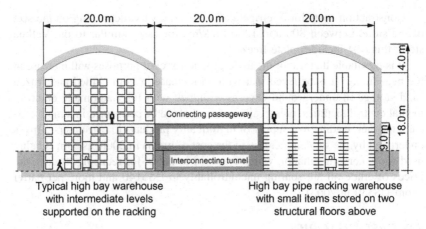

FIGURE 5.7 Schematic cross section of typical warehouse caverns [based on [15]].

linked by a truck-docking cavern that has a span of 18 m, a height of 8.4 m, and a length of 210 m.

Facilities include an access tunnel 45 m² in cross section and 225 m in length, at a gradient of 1:8, leading from the portal down to the truck-docking cavern; an air supply and emergency escape tunnel with a cross section of 15 m²; and a total of 10 interconnecting tunnels of 12 m² cross section between the warehouse caverns. The total excavated volume of the warehouse complex is approximately 270,000 m³, providing a storage volume of some 210,000 m³. As illustrated in the typical cross section shown in Figure 5.7, the warehouse will be on two or three levels in the caverns. The proposed main storage system is high-bay warehousing, with 9-m floor-to-floor heights, employing narrow-aisle side loaders to access pallets stacked high (Fig. 5.7). Vertical transport is by means of heavy-duty hoists at the loading bay end of each cavern [14].

Parts of the cavern volume are planned to be for storage of small items and archives, supported on structural floors. In the cost estimate, two structural floors have been included for each cavern to cover a variety of cavern utilization options; allowance is made for air-conditioned office space within the cavern complex.

The design of the warehouse complex allows part or all of the cavern volume to be used for chilled and humidity-controlled storage.

5.9.2 Cold Storage Plant in Bergen, Norway [16]

Cold storage is located outside Bergen, Norway, where the climate is fairly mild and wet. The plant is situated at the foot of a steep mountain at an elevation of 47 m. There are almost no overburden deposits, and the bedrock consists of Precambrian granitic gneiss. To establish an area outside the storage room, some

20,000 m³ rock was excavated, with a 23-m-high cut at the most. The work was completed in 1974 and the facility has been under operation since then.

The underground cold storage facility consists of a 12-m-long access tunnel and a storage cavern 57 m long and 20 m wide, with a total height of 10.8 m. The volume of the storage room is approximately 11,000 m³. The outline of the cavern is shown in Figure 5.8. The rock surface is exposed in the cavern roof and walls, and 3-m-long grouted rock bolts (one bolt per 5 m²) are the only rock-supporting means used. A concrete wall with gates is installed at the entrance. The concrete floor has rails on which wheeled racks can move, thus allowing maximum utilization of the volume; containers are moved to and from the racks by electrical trucks.

The machine is situated inside the rock cavern, whereas the office, service buildings, and ventilation equipment are placed outside of the cavern. The cavern is cooled by two electric compressors of 72 Hp each. The temperature was initially −28°C and was later raised to −22°C, which is now the operating temperature.

Water seepage occurred as dripping at some 15–20 spots. During the cooling-down period, most water seepage disappeared quickly, but seepage increased at 2–3 spots, and for about 3 weeks they were busy with picking and removing ice. Some holes were drilled to collect the water at concentrated spots. Fans blowing cold air were placed on these spots and the water froze rather quickly. The air temperature in the cavern was then −14°C.

The time–temperature diagram during the cooling period is shown in Figure 5.9. The temperature decreased rapidly to −10°C. From 10° to −15°C the fall in temperature was rather slow, probably due to water freezing

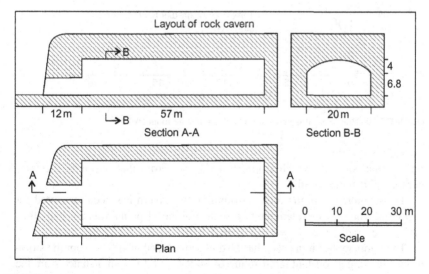

FIGURE 5.8 Layout and cross section of cold storage plant in Bergen, Norway [16].

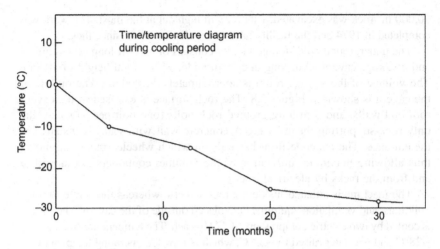

FIGURE 5.9   Variation of temperature with time during cooling [16].

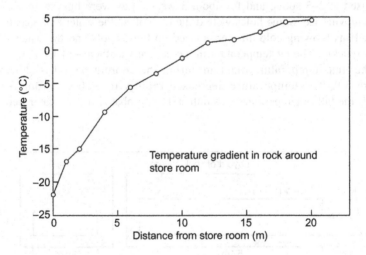

FIGURE 5.10   Variation of temperature in rock from store room [16].

on the rock surface, which stopped at −14°C. From that moment, the temperature fell more rapidly.

The temperature in the rock surrounding the cavern has been measured for 3 months. The measured temperature at each of the 11 points was constant, thus giving the gradient as shown in Figure 5.10.

The temperature in the rock has also been recorded at 100 m from the storeroom, showing a constant temperature of +4.9°C at 6 to 10 m distance from the terrain surface.

REFERENCES

[1] Dunkel FV. A new era for underground food storage. Tunnel Undergr Space Technol 1987;2(4):365.

[2] Dunkel FV. Applying current technologies to large-scale, underground grain storage. Tunnel Undergr Space Technol 1995;10(4); pp. 477–96.

[3] Dunkel FV, Sterling RL, Meixel GD, Bullerman LB. Underground structures for grain storage: interrelatedness of biologic, thermal and economic aspects. In: Hou XY, editor. Advances in geotecturol design. Shanghai: Tongji University Press; 1988. p. 1–8.

[4] Dunkel FV, Lareen R, Christensen M. Use of a two-dimensional model to detect the most biologically and economically feasible underground grain storage structure. In: Bartali H, editor. Proceedings of the international seminar on storage structures for grains, legumes and their derivatives. Rabat, Morocco: International Commission on Agricultural Engineering; 1990. p. 439–50.

[5] Maier DE. Chilled aeration and storage of U.S. crops: a review. In: Proceedings of the sixth international working conference on stored product protection. Canberra, Australia; CAB International, Wallingford, United Kingdom; 1994. pp. 300–311.

[6] Bailey JE. Whole grain storage. In: Christiansen CM, editor. Storage of cereal grain and their products. St. Paul, MN: American Association of Cereal Chemists; 1974; pp. 53–78.

[7] Booth RH, Shaw RL. Principles of potato storage, International Potato Centre, Lima, Peru, in paper on "Underground Storage of Food" by Raymond R. Sterling, George D. Meixel, Florence Dunkel, and Charles Fairhurst. Undergr Space 1981;7: pp. 257–62.

[8] Ikaheimonen P, Leinonen J, Marjosalmi J, Paavola P, Saari K, Salonen A, Savolainen I. Underground storage facilities in Finland. Tunnel Undergr Space Technol 1989;4(1): pp. 11–15.

[9] Zhao J, Choa V, Broms BB. Construction and utilization of rock caverns in Singapore, part B: development costs and utilization. Tunnel Undergr Space Technol 1996;11: pp. 73–79.

[10] Zheng SC. The effect of thermal insulation and moisture control in underground cold storage. Tunnel Undergr Space Technol 1989;4: pp. 503–7.

[11] Broch E. Use, planning and design of underground structures. In: Proceedings of the symposium on underground space and construction technology. Seoul, Korea; 1993. pp. 31–47.

[12] Unver B, Agan C. Application of heat transfer analysis for frozen food storage caverns. Tunnel Undergr Space Technol 2003;18: pp. 7–17.

[13] Broch E, Frivik PE, Dorum M. Storing of food and drinking water in rock caverns in Norway. In: Proceedings of '94 international symposium for grain elevator and underground food storage. Seoul, Korea; 1994. pp. 261–317.

[14] Wallace JC, Ho CE, Bergh-Chrlatensen J, Zhao J, Zhou YX, Choa V. A proposed warehouse-shelter cavern scheme in Singapore granite. Tunnel Undergr Space Technol 1995;10(2): pp. 63–67.

[15] Littlechild BD, Kiaerstad O, Harley MV, Goldstein AL. Two warehouse schemes in Hong Kong. Proc. Seminar, "Rock Cavern–Hong Kong." Hong Kong: Institution of Mining and Metallurgy, Hong Kong Section; 1989. pp. 109–27.

[16] Barbo TF, Bollingno P. Experience from cold storage plant in rock cavern. Norwegian Hard Rock Tunnelling. University of Trondheim, Norway: Norwegian Rock and Soil Engineering Association, Publication No. 1; 1982. pp. 47–49.

Chapter 6

Underground Storage of Water

An investment in knowledge always pays the best interest.

Benjamin Franklin

6.1 GENERAL

Availability and access to fresh and safe drinking water is a precondition for life. Freshwater resources are rainwater collected on the ground and groundwater collected at various depths below the ground surface. These resources are, however, often insufficient or too costly for basic water requirements in many parts of the world. There is an acute shortage of drinking water in developing and poor nations, especially in rocky areas and hilly terrains where it doesn't rain as much. Safe drinking water is going to be a vast business opportunity all over the world.

The traditional way of ensuring a constant water supply, however, is collecting and storing water in surface dams. In open reservoirs, drinking water is exposed to the influence of sunlight and pollution from the air. In arid regions, a great proportion of water evaporates from surface open reservoirs. Moreover, open reservoirs are commonly situated in natural or artificial depressions and will therefore tend to collect drainage from the surrounding landscape. If such reservoirs are located close to populated areas, there is a danger that polluted surface water or groundwater may be drained into the drinking water.

Today, open reservoirs for the storage of drinking water are not normally acceptable to the health authorities in Norway. They have to construct closed tanks of some kind, and old schemes with open reservoirs often have to be redesigned and reconstructed.

Recharging of groundwater may eliminate wide cracks and subsidence in clayey agricultural fields due to excessive pumping out of the groundwater, thereby preventing damages to the fields.

In principle, there are four ways of recharging and storing water underground:

- Recharge into nonconsolidated formations such as porous alluvial material, including river beds, alluvial fans, and suitable aquifers.
- Recharge into consolidated formations such as porous (karstified) limestone and sandstone aquifers.

89

- Recharge into crystallized rock formations.
- Storage in man-made caverns or reservoirs.

6.2 WATER STORAGE BY RECHARGE METHODS

There are four basic ways, as given previously, by which water can be stored underground using the recharge methods. The method and its controlling factors are given in the following paragraphs [1].

6.2.1 Controlling Factors

The most important factors regulating the storage of water underground are the porosity and permeability of a geological formation. The effective porosity of a sandstone formation is normally 10 to 20% and the figure for alluvial sand and gravel deposits is about the same. This means that in a river bed 200 m wide and 1000 m long, groundwater between 0.2 and 2.0 MCM can be stored if the saturated depth is 10 m. This is enough water supply for approximately 3000–30,000 people. In terms of agriculture, approximately 4–400 hectares can be irrigated with this amount of water (for crops under winter conditions.).

The usually heavy load of sediment such as silt and clay, which is carried by surface water, tends to clog the surface of infiltration works and stops recharging to the underground facility. Therefore, the water must be treated before entering infiltration works or infiltrated through a geological formation so that it will not be affected by river sediments. Concentrated recharge or infiltration to geological formations with small pores (e.g., sandstone aquifers) therefore requires special techniques.

However, in fractured aquifers such as karstified limestone and dolomite formations, water is usually stored in and moves along few, but large, open fractures, faults, and solution cavities. Here the water can be recharged through a well if the well has penetrated through enough open fractures.

In hard rock formations (e.g., crystalline rock such as granite), research in Sweden has shown that underground dams can be constructed to effectively seal off groundwater outflow from a specific basin or catchment area [2]. Losses from such a groundwater system are thus reduced greatly; some 50,000–100,000 m^3/km^2 of water can be stored and utilized.

6.2.2 Recharge Methods

Surface water can be added to a geological formation through surface infiltration, infiltration wells, or infiltration galleries.

The governing factors for surface infiltration are the amount of suspended solids in the infiltration water and the area for infiltration. The more important of these two factors is the sediment load. If this is high, it will quickly close the open pores in the formation and infiltration will stop. As the sediment load is very high in arid areas, due to the rapid flow of water, infiltration has to include ways of reducing the sediment load before infiltration into the ground can take place.

This may be achieved, for example, by installing a drainage system at the bottom of the dam reservoir. After filling the dam, the runoff water will be discharged through this drainage system. In most cases it will be free from impurities and can be either used for direct consumption or recharged into the groundwater system through wells, galleries, and so on for later use. Another method is to store the runoff for a sufficiently long time to let most of the sediment settle in the dam. The water can then be siphoned off and recharged at the desired place. Evaporation losses will probably be larger, however, with this system.

One method requiring a minimum of technical measures is storing surface water in river bed (wadi) material. Most river beds consist of porous sand and gravel. Water stored in this material creates a groundwater body that can be utilized for agriculture and domestic consumption. One way of doing this is through the use of underground dams (Fig. 6.1). Several underground bodies of water can be created to prevent normal outflow losses. The recharge to these underground dams is mainly through direct infiltration during times of river flow. Unfortunately, most of the surface water will still be lost. However, once stored in the ground, the water will not be exposed to direct evaporation and loss will be negligible.

A completely different approach to the river bed infiltration technique is the construction or creation of new reservoirs of river bed material at selected sites where suitable conditions did not exist previously. This is achieved by storage dam in stages (Fig. 6.2). The dam collects the coarser material transported by the runoff water while permitting finer material to pass through. The reservoir volume is filled quickly, and the water is stored in the intergranular space in the sand and gravel. A second stage of the dam is then constructed and the reservoir is enlarged further. More stages can be built depending on the topographical conditions, water requirements, etc. [2]. Groundwater in this artificial dam is

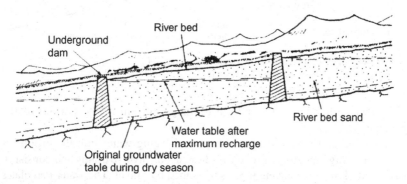

FIGURE 6.1 Impoundment of water by means of underground dams [1].

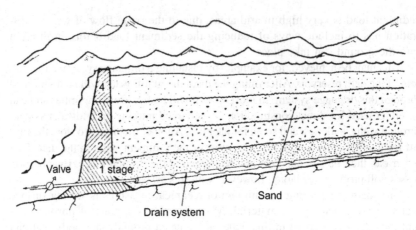

FIGURE 6.2 General scheme for a multistage dam reservoir [1].

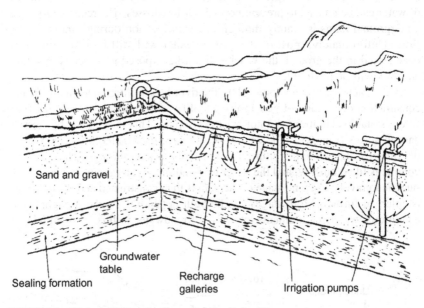

FIGURE 6.3 Method of recharging by means of galleries [1].

withdrawn either by wells or by a drainage system in the bottom of the reservoir.

A second and very efficient way of recharging water into the ground is through recharge galleries (Fig. 6.3). These act as inverse drains and consist of perforated plastic or concrete pipes surrounded by gravel. The water percolates out through the pipe and gravel into the formation, where it then seeps downward. The advantage of this system, as compared to the first, is that loss from evapotranspiration is reduced and that irrigation methods using less water can be applied. Because of the intermittent character of the recharge process, such galleries are not clogged as easily as infiltration basins.

A third method of introducing surface water into the ground is through infiltration wells (Fig. 6.4). These, however, require very clean water to be recharged into a porous formation, and it is unlikely that such wells can be used for the infiltration of mud-loaded runoff water. However, the method can be used when water is recharged into a karstified formation, as the open pores here consist of solution cavities.

If there are no sufficient suitable geological formations for infiltration or storage, a dam reservoir with connecting rock tunnels or large pipes can be used to transport the water to more suitable cities. A continuous system of rock tunnels or pipes eases collection of the surface water and transport of water to suitable underground storage facilities.

The isotope technique may be used effectively to identify recharge areas and construction of water conservation and recharge structures such as dikes, check dams, and contour trenches in dry hilly areas for controlling subsurface flow, rainwater harvesting, and groundwater harvesting.

A number of underground openings have been excavated in the hard and strong rock masses of Norway as replacements for or alternatives to open reservoirs, concrete tanks, or steel tanks for the storage of drinking water [3]. A closed water tank, which is what a rock cavern tank is, has several advantages over traditional open reservoirs. Above all, it is easier to keep pollution under control with a closed tank. Mostly the work of Broch and Odegaard [3] has been referred to for the preparation of this chapter.

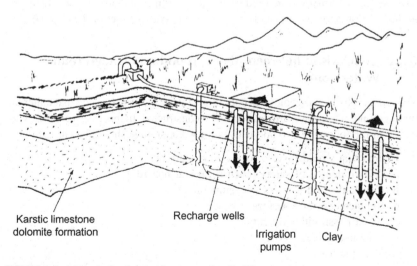

Karstic limestone dolomite formation

Recharge wells

Irrigation pumps

Clay

FIGURE 6.4 Method of recharging by means of wells [1].

6.3 UNDERGROUND ROCK CAVERN TANK STORAGE

6.3.1 Function and Location of Water Tanks

The basic function of a water tank is to act as a storage buffer to meet variations in consumption and keep the water head stable. This makes it easier to operate the treatment plant and the pumps at constant capacities and allows smaller dimensions of the main pipe lines. The water tank also acts as emergency storage in case of fire or failure in the supply system. Swamee and Sharma [4] have described a computer program for the design of water network systems.

Small water tanks are normally single chamber tanks. For volumes exceeding approximately 10,000 m³, double or even multiple chamber tanks are often used. This allows one chamber to be emptied for cleaning and maintenance without interrupting the water supply.

A water tank should be situated at an elevation that gives a suitable water pressure in the consumption area. It is also preferable to locate the tank as close to the consumption area as possible, especially if the capacity of the tank is designed to cover the variations in daily consumption.

Most water tanks in Norway have been freestanding structures made of conventional reinforced concrete or prestressed concrete. When double chambers were necessary, either two separate structures were constructed or two concentric chambers were built in one structure. To minimize the impact of such concrete structures on the environment, they have often been dug partially into the ground or hidden in other ways, especially in hill tops.

One way of making water tanks invisible, of course, is to put the tank completely underground. The most economical way of doing this is to use the rock mass as the construction material, that is, excavating caverns for water in bedrock. A rock cavern tank is normally composed of an access tunnel and one or more chambers in the form of large, short tunnels; there is a dam wall in front of the chamber.

6.3.2 Comparison between Aboveground and Underground Water Tanks

The topographical and geological conditions in an area may be such that either an underground or a surface structure could be constructed. In this case a careful evaluation of the following two alternatives for water storage should be carried out.

1. Factors that would favor a rock cavern tank are:
   a. A high degree of safety (even from war hazards)
   b. Constant—and low—water temperature
   c. Replacement of a surface eyesore
   d. Good possibilities for future extension
   e. Low maintenance costs
   f. Multiple uses
   g. Little or no additional cost for a two-chamber facility

Factors unfavorable to an underground cavern tank are that:
- Polluted groundwater may seep into the drinking water.
- Water may leak out through the rock mass.

2. Factors favoring a surface or overhead tank, however, include:
   a. Elimination of the risk of polluting the water (except in the case of sabotage)
   b. Ease of leakage detection and repair

The following disadvantages with a surface facility may be mentioned:
   c. The surface tank occupies building ground.
   d. The tank may be regarded as an eyesore and thus resisted by the neighborhood.
   e. The tank may endanger the surrounding area in case of war.
   f. The water temperature will change with seasons.

Even though the aforementioned positive and negative factors are important enough in themselves, the decisive factor for choosing the type of water tank will normally be cost.

6.3.3 Planning and Design

As the rock cavern tank is basically a rock mass structure, the success of planning and design depends to a large extent on cooperation between the consulting engineer and the engineering geologist. The general design procedure for tunnels and caverns in Norway is divided into four stages [5,6].

1. A location is selected from a stability point of view that shows the optimal engineering geological conditions of the area.
2. The length axis of tunnels and caverns is oriented so as to give minimal stability problems and overbreak.
3. The shape of caverns and tunnels takes into account the mechanical properties and the jointing of the rock mass as well as local stress conditions.
4. The different parts of the total complex are dimensioned so as to give an optimal economic solution.

Mistakes in any one stage will result in overall economic consequences, the extent of which will vary with local conditions and the type of project. Normally, rock cavern tanks in Norway are unlined and have a limited overburden. This implies that the rock mass is subject to low stresses. The joints may thus be more open than joints at a deeper level in the rock mass. It is therefore important that the permeability (k) of the rock mass, that is, the possibility of leakage along intersecting joint sets, be taken into consideration when water tanks in rock are designed ($k \cong 100/Q.q_c$ lugeons). Generally, stiff rocks, such as granites and quartzites, have a tendency to greater leakage than more deformable rocks such as micaschists and phyllites due to clay coating of joints, etc. Carbonate rocks (limestones and marbles) and rock masses with calcite-containing joints and faults are of special interest from a leakage point of view as calcite is dissolved easily by water.

The most important decision to be made during planning of a rock cavern tank is the location of the tank. It should not be forgotten that when the location of an underground opening is decided, the choice of material into which the opening is going to be excavated is also made. It is therefore of the utmost importance that this crucial decision be based on the advice of an experienced engineering geologist.

The engineering geologist will have to carry out geological and geotechnical investigations of alternative sites to give this advice. At this stage, it is also of particular interest to get information about rock types and weakness zones (or faults). Combined with information from the consulting engineer about the upper and lower water levels and the approximate capacity of the tank, and with information about the topography from detailed maps and air photos, the engineering geologist will be able to eliminate unfavorable rock sites, finally ending up with a limited number of possible sites.

Before the consulting engineer starts planning possible layouts of the underground facility within these areas, an alignment of the (longer) axis of caverns that will yield the best stability and least leakage should be known. Evaluations of joint sets in the rock mass by the engineering geologist will provide the answer. This information is also important when the shape and the dimensions of the different parts of the rock caverns and the connecting tunnels are to be decided.

A particular problem for rock cavern tanks (and for all other underground designs) is the entrance (portal). This is the only part that will be visible to the public. From an excavation point of view, it is one of the most difficult parts, as the rock mass is generally unstable due to slope and weathering. First of all, great care should be taken to find the most suitable site for a stable portal. As for the quality of excavation, restrictions should be put on the contractor's work. All too often one can find ugly tunnel entrances where the rock mass has been torn up unnecessarily by too hard blasting. A combination of knowledge of the rock mass and careful blasting is the only way to get a proper result. A thorough discussion of the excavation of tunnel portals is given by [7].

6.3.4 Cost for Underground Cavern Storage

The costs for a number of conventional water tanks of reinforced concrete and for rock cavern tanks have been converted to the 1979 Norwegian price level for the purpose of comparison. Results are summarized in Figure 6.5 (prices are also given in U.S. dollars). The two curves intersect at a storage volume of about 8000 m^3, beyond which a rock cavern tank will normally be a less expensive solution in hard rocks in Norway.

Poor rock conditions will increase the costs of support in unlined rock caverns and thus show an intersection point higher than 8000 m^3. However, if the excavated rock can be sold or if the price of land for freestanding

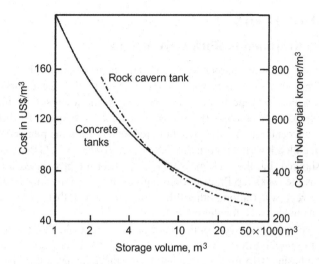

FIGURE 6.5 Specific costs per cubic meter of storage volume for conventional water storage tanks of reinforced concrete and for unlined rock cavern tanks [3].

water tanks is high, this will favor the choice of rock cavern tanks for storage volumes of even less than 8000 m³.

6.3.5 Construction and Maintenance Experience

In the entrance tunnel with its cold pipes and valves in cold climates, it is important to reduce the moisture content in the air by sufficient ventilation. Leakage from the rock may be drained through perforated plastic tubes before shotcreting is done. In a cavern storage facility at Steinan, a total of 500 m of such tubes were installed in the roof and walls. The tubes emptied into draining trenches along the walls for proper drainage of rock mass.

A careful registration of all leakage during the excavation period is important for successful drainage in the service and entrance tunnel, as also in the water basins. Open joints will then be observed and can be sealed or grouted before the basins are filled with water. Because of drawdown in the groundwater table above the rock caverns, water will normally flow toward the basins (tanks) rather than from them. Leakage from water tanks in rock has seldom been observed, nor has seepage of polluted water into these tanks been reported. One should always be, however, aware of these possibilities.

All drinking water tanks should be emptied, inspected for leakage, cleaned, and disinfected regularly. To facilitate cleaning, tubes with valves at intervals of about 12 m should be installed along one of the walls in the basins. If the valves are left open when the basin (tank) is filled with water, the same tube-and-valve system may be used to create circulation in the stored water when this is necessary.

6.4 CASE HISTORIES

6.4.1 The Kvernberget Rock Cavern Tank

Kristiansund is a fishing harbor located on an island off the northwest coast of Norway. The need for a new water supply system made it necessary to cross two fjords with a pipeline from an inland lake. At the town side of the fjords, a water reservoir situated 80–100 m above sea level was needed. About 2 km from the town center, and only some hundred meters from the planned pipeline, a mountain called Kvernberget rises to 200 m above level. Geological investigations showed that the rock mass of Kvernberget might be used for a water reservoir, as the rocks are Precambrian gneiss. The decision to locate the water reservoir in rock was (among other things) due to the fact that future expansion could be planned easily for the reservoir.

Figure 6.6 shows the layout of the Kvernberget water tank. Two basins of 11 × 7.5 × 120 m give an effective volume of 8000 m^3 of water each. The distance between the basins is 15 m. By extending the entrance tunnel, a third basin could be excavated without disturbing the operation of the two existing basins.

Along the outer part of the entrance tunnel, a service section contains all the equipment for operation of the whole water supply system. To support the rock in the basins, 100 m^2 of shotcrete was applied and about 100 rock bolts were installed. The rock was cleaned thoroughly, and the basins (tanks) have a concrete floor. The outer part of the entrance tunnel (portal) is fully shotcreted with 1200 m^2 of shotcrete.

No leakage was observed when the basins were first filled with water and, after 3 years in operation, no water loss has been observed.

A total volume of 21,000 m^3 of solid rock was excavated at a price of 2.95 million Norwegian kroner (Nkr) (cost of transportation of muck and support of the rock mass included). This gives a price of 140 Nkr per m^3.

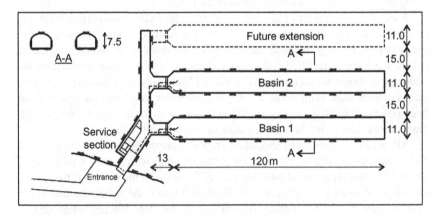

FIGURE 6.6 Layout of the Kvernberget rock cavern tank in Norway [3].

Total costs for the water reservoir, including the central operation, were 5.05 million NKr (1 U.S. dollar = 5 Nkr).

Only a year before the Kvernberget water tank in rock was completed, a conventional "above-the-ground" water tank with exactly the same capacity, 16,000 m³, was put into operation in Huseby in the city of Trondhiem, Norway. Total cost for this tank (cost for building site excluded) was 5.9 million N.kr (all prices based on 1979 levels). Thus a large underground rock tank is more economical than that aboveground.

6.4.2 The Steinan Rock Cavern Tank, Norway

Another water tank, the Steinan rock cavern tank, was put into operation in Trondheim. The 20,000-m³ capacity cavern started operation in 1979. As can be seen in Figure 6.7, the layout is very much the same as the Kvernberget tank (Fig. 6.6), with two basins and a service section close to the entrance. The reason for showing the Steinan tank is to demonstrate that even where the geological conditions are not as simple and favorable as at Kvernberget, a rock cavern tank is a possible and economical solution.

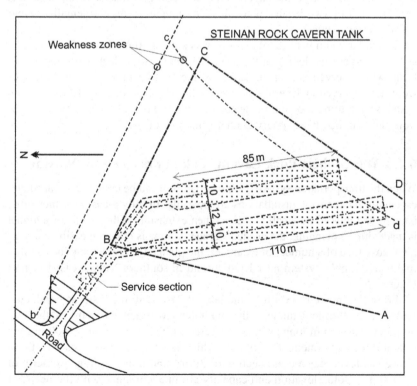

FIGURE 6.7 Layout of the Steinan rock cavern tank; storage capacity 20,000 m³ [8].

At Steinan, the water level for hydraulic reasons would have to be situated at 190 ± 10 m above mean sea level. The road level is around 185 m. Based on a study of the geological conditions and the desired size and shape of the basins, a minimum rock mass overburden was defined. A line A–B (Fig. 6.7) could thus be drawn along the 215-m above mean sea level curve. This defined the outer limits in the hillside for the basins.

Two weakness zones were outcropping in the hill along the dotted lines b–c and c–d (Fig. 6.7). Strikes and dips of these were measured, and the zones were projected down to the level of the basins (tanks). To account for uncertainties in the measurements, safety margins of 10 m were added and lines B–C and C–D were drawn (Fig. 6.7). The area where the basins could be placed in undisturbed, stable rock masses was now limited by the lines A-B–C-D with a rock cover of about 30 m and more.

To find the most stable orientation for the long caverns in the highly deformed greenstone pillow lava, a thorough survey of the joint sets was carried out. The dominating steep joint sets had strikes of N70-80°W and N75-90°E. According to orientation rules (see Section 4.4), the (longer) axis of the caverns was oriented along the direction N10°W so that the strike of the joint plane is nearly perpendicular to the cavern axis. The desired capacity of the tank was thus oriented by two basins of width 12 m, height 10 m, and lengths 90 and 115 m, respectively.

The pillar width between the basins was kept at 10 m, that is, 5 m less than at Kvernberget (just less than the span or height). Some leakage in the empty basin was observed during the filling of the first basin at Steinan. Leakages were not observed at Kvernberget, as such a distance of only 10 m between basins seems to be too small if leakages have to be avoided [8]. No leakages from the tank had been observed after 6 months of operation.

6.4.3 The Groheia Rock Cavern Tank in Kristiansand, Norway

When the transportation system itself or a part of it can be used as a storage tank, especially inexpensive installations are possible. In hilly areas, pipelines often have to be replaced by small tunnels. When extension of the profile of a tunnel is made during planning, the needed storage volume is obtained easily at a marginal cost. One of a number of examples, the planned Tronstadvann intermunicipal water supply system near Kristiansand in southern Norway, is shown in Figure 6.8.

From the intake reservoir of the lake of Tronstad, the water is conducted through a 3400-m-long tunnel with a minimum cross section of 8 m^2. At the lowest point of a 3900-m-long pipeline the water is lifted by pumps to a 1600-m-long tunnel through Groheia. This tunnel, which is situated approximately 100 m above sea level, has a cross section of 30 m^2 and thus a storage capacity of 48,000 m^3. Both elevation and capacity are in accordance with the needs of the scheme. Rocks in the area are sound Precambrian gneisses. A few weakness

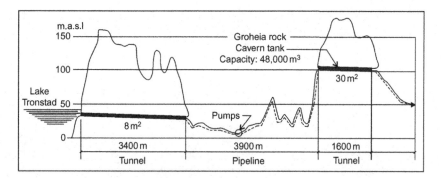

FIGURE 6.8   The Groheia rock cavern tank, part of the Tronstadvann intermunicipal water supply system [3].

zones cross the tunnel and some support measures may be needed. The tunnel is, however, unlined like the aforementioned rock tanks.

REFERENCES

[1] Moller A. Underground storage of water. Undergr Space 1983;7:264–6.
[2] Nilsson A. Appropriate technology for artificial sub-surface storage of water in developing countries. Stockholm (Sweden): Department of Land Improvement and Drainage, Royal Institute of Technology; 1982.
[3] Broch E, Odegaard L. Storing water in rock caverns. Undergr Space 1983;7:269–72.
[4] Swamee PK, Sharma AK. Design of water supply pipe network. Hoboken (NJ): Wiley Interscience; 2008. p. 353.
[5] Broch E, Rygh J. Permanent underground openings in Norway: design approach and some examples. Undergr Space 1976;1(2):87–100.
[6] Selmer-Olsen R, Broch E. General design procedure for underground openings in Norway. In: Bergman SM, editor. Storage in excavated rock caverns, vol. 2. Oxford: Pergamon Press; 1977. p. 219–26.
[7] Garshol K. Complicated portal blasting, principles and methods based on experience. In: Proceedings of the 4th International Congress, vol. 1. Montreux, Rotterdam: International Society for Rock Mechanics, Balkema; 1979. p. 393–9.
[8] Broch E, Odegaard L. Storing water in rock caverns is safe and cheap. Norwegian Hard Rock Tunnelling, publication No. 1, Norwegian Rock and Soil Engineering Association, University of Trondheim, Norway; 1982. p. 69–76.

Chapter 7

Underground Parking

You are the creator of your own destiny.

Swami Vivekananda

7.1 INTRODUCTION

The progressive increase of the number of cars parked on the roadside and often on pavements has caused a significant decrease of the road space available to both moving vehicles and pedestrians.

The lack of sufficient parking space in the densely inhabited city center of Athens is a serious problem that degrades the quality of life of the inhabitants. The growing number of automobiles in Japan has caused serious parking problems, such as a lack of parking spaces and illegal parking on public roads, especially in the city centers. Construction of underground parking facilities has increasingly drawn attention as a solution for parking problems in urban areas. To date, underground parking has been constructed in conjunction with underground shopping malls. However, underground spaces used specifically or mainly for parking have been proposed by several companies [1]. Many affluent countries have adopted the option of underground car parks to overcome this problem.

7.2 TYPES OF PARKING FACILITIES

Parking facilities may be divided into the following three groups, based on their location in relation to the surface [2]:

1. Surface (flat) structure
2. Aboveground (elevated) structure (Fig. 7.1)
3. Underground structure

The latter two types of parking structures may be further categorized as mechanized or nonmechanized, depending on whether the vehicle is moved into its parking space by mechanical means or by its own power. Automatic or mechanical parking systems are about making the best use of available space above and below ground. With less environmental impact, reduced opportunities for theft and vandalism, and real cost benefits, automatic parking is the new watchword in urban planning.

Underground Infrastructures

FIGURE 7.1   Multi-story parking facility on surface. *(source: www.sachinwalvekar.wordpress.com)*

In contrast to the conventional self-driving system, which allows drivers to move their cars to the parking position, mechanically driven parking systems relocate a car from the entrance to the parking position automatically. Experience has shown that the parking space requirement for an automatic mechanical car store is roughly one-third to one-half of the space requirements for a conventional park equipped with a ramp. Furthermore, the risk of accidents is almost nonexistent because the store is unmanned except for inspection and service work [3].

Viewed historically, the oldest form of vehicle parking is street parking, followed by multilevel elevated facilities and, more recently, underground facilities, which originally were conceived in pursuit of more efficient land use. Since the introduction of mechanized parking, efficiency through multilevel construction has been reemphasized (Fig. 7.2). Underground parking is popular in Montreal, Canada.

Although it is desirable that parking facilities be constructed in or as close as possible to the zone in which the demand originates (e.g., near underground metro stations), this goal has become difficult to achieve sometimes. Accordingly, parking facilities aboveground and/or underground are being constructed in conjunction with large buildings that are part of urban redevelopment projects. If such an arrangement is not possible, usable space must be sought underground, beneath public land such as street and parks. Further, because the space at shallow depths below streets is almost always occupied by existing facilities, such as subways or utility tunnels, parking facilities constructed later must be built to a deeper level, despite the resulting increase in construction costs. A comparison of the characteristics of the various locational types of parking facilities is shown in

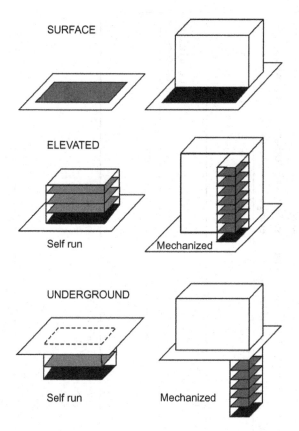

FIGURE 7.2 Structural type of parking facilities [2].

Table 7.1. The descriptive items in Table 7.1 have been evaluated by inference from numerous examples of public structures in urban areas and related information and are not definitive. With respect to the quantifiable items, underground facilities are clearly at a disadvantage in comparison with the other two types. Where is the land for surface parking in mega cities? Nonetheless, the share of underground parking facilities will probably grow to exceed that of other types in the future, for three main reasons [2].

1. In larger cities, it has become difficult to find surface space for parking facilities.
2. Given the growing public concern regarding urban aesthetics and surface harmony, powerful resistance is expected to increase against new construction of surface and aboveground parking facilities.
3. An underground parking facility is safe during a major earthquake, whereas a building for parking aboveground needs to be designed to be safe during a major earthquake.

TABLE 7.1 Characteristics of Parking Facility by Location in Relation to the Surface [5]

Locational Type of Parking Facility	Quantifiable Factor					Nonquantifiable Factor		
	Construction Cost	Required Floor Area[a] (m²/vehicle)	Required Surface Area[a] (m²/vehicle)	Operating Expense	Ease of Converting/ Rebuilding Facility	Users' Psychological Feelings	Urban Aesthetics	Annoyance to Environment
Surface	Slight	20–30	20–30		Free		Poor	Disagreeable (engine noise and exhaust gases)
Aboveground (non-mechanized)	Medium	25–35	5–7 (five tier)	Energy for lighting	Difficult		Poor	Disagreeable (engine noise and exhaust gases)
Aboveground (mechanized)	Large	15	3–5	Maintenance cost for transfer machines	Difficult		Poor	
Underground (non-mechanized)	Large	30–45	0	Energy for lighting; maintenance cost for transfer machines	Impossible	Uneasy		
Underground (mechanized)	Huge	15	0	Maintenance cost for transfer machines	Impossible			

[a]Based on guidelines of the Japan Society of Parking Engineering.

7.3 VARIOUS MODERN MECHANICAL UNDERGROUND PARKING OPTIONS

7.3.1 Trevipark System

Traffic congestion is a major problem in all mega cities. Even when actively promoting public transport services, vehicles still overpower the stretched road systems. Land in city centers commands such premium prices that car parking land is becoming less available. However, an ingenious new automated system could change the situation. The late 1990s saw the first commercial installation of the Trevipark in Cesena, Italy (www.trevipark.co.uk). The Trevipark system solves many of the traditional problems associated with urban parking—congestion, pollution, land space, and security—through the installation of compact, circular, underground silos that optimize space, are installed easily, are secure, and are completely automatic. A Trevipark is currently being installed throughout Turin, Italy.

Trevipark offers cost-effective, fully automatic, unmanned underground or aboveground car parking system solutions, particularly for tight sites in urban areas where space is at a premium. The car park is fully computer controlled from within the silo, requiring no permanent operatives or attendants, and conforms fully to ISO 9001/2 standards and European safety regulations for the following reasons:

- The driver and passengers alight from the car, leaving it in the entry bay with the engine off (reducing emissions).
- There is no need for ramps or driveways.
- The car is parked automatically without anybody entering the car park.
- The car is returned to the exit bay, facing the right direction, ready to be driven away.
- Once parked, the cars are totally safe and secure from being stolen, vandalized, or damaged.
- Assistance can be obtained via a telephone link.
- A management control center operated by a facility management company, or in-house with training given, monitors the car park by closed circuit television (CCTV).
- The command center can override the onsite computer, should the need arise. One command center can control many independent systems.
- Parking and retrieval times are identical, averaging 50 seconds per car.
- Tariff charges can be payment-based using cash, credit, or dedicated card.
- The system is thoroughly reliable because all essential plants and equipment have a backup facility and are monitored 24 hours a day, 7 days a week by a computer.

A standard silo can accommodate a maximum of 108 cars in 441 m² (ground surface area), that is, 21 × 21 m; the car park can be built on completion using the car parking structure as part of the foundations. Construction of a standard Trevipark takes between 5 and 6 months, depending on site conditions and access facility. The entry and exit positions are normally separate, where space will allow, but if there is not sufficient space these can be combined. The computerized system ensures that cars are parked in the nearest available space. The Trevipark system can be custom designed to suit individual sites. Figures 7.3, 7.4, and 7.5 show three Trevipark systems. Figure 7.6 gives the inner view of the same.

In Europe, 1200 Trevipark car spaces have been constructed and over 2000 spaces are in the course of construction or negotiation. Since opening the first public car park in Cesena, Italy, Trevipark has generated both national and worldwide interest in its systems.

7.3.2 Douskos Car Parks—Mechanical Underground Parking Station Systems

The invention refers to mechanical underground car parks and parking station systems (www.parkingstation.com) that are constructed under the surface of streets, pedestrian areas, and other public areas. Existing other mechanical car parks, automated parking station systems, and those of traditional type, nonmechanical parking stations are constructed underground or overground in public or private areas. These car parks and parking station systems occupy

FIGURE 7.3 Trevipark system with 47 car capacity; 23 aboveground and 24 underground. *(source: www.trevipark.co.uk)*

FIGURE 7.4 Trevipark system with indirect loading bay for 72 cars underground. *(source: www .trevipark.co.uk)*

FIGURE 7.5 Trevipark system with direct loading bay for 84-car parking capacity underground. *(source: www.trevipark.co.uk)*

FIGURE 7.6   Inner view of Trevipark system. *(source: www.trevipark.co.uk)*

large areas of land belonging to the state or private companies, waste a lot of space (30 m^2 for a car instead of only 12 m^2 in the Douskos system), and are not usually situated near the driver's place of work or home. They have inadequate capacity for a minimum percentage of cars; the result of this is the continuation of illegal parking on the streets and the hindrance of traffic. The price of land for these car parks and parking station systems require extremely large amounts of money to be invested, a fact that makes them disadvantageous.

The advantage of this invention (Fig. 7.7) is that it is constructed by the state or the municipality or by self-finance methods or private companies and ensures

1 Car is entering the parking.
2 Moves on the rails inside the tunnel.
3 Is placed upon the first available parking place.
* For exiting the car park, the reversed procedure is followed. Copyright © 2005 Douskos Car Parks

FIGURE 7.7 Douskos mechanical underground parking station system. *(source: www.parkingstation .com)*

that most of the residents of every house or apartment will have permanent parking spaces in front of their houses.

Advantages of Douskos Mechanical Parking System

Some of the advantages from economical, environmental, aesthetical, and traffic circulation importance points of view that result from the manufacture of "Mechanical Underground Car Parks under Streets and Pedestrian Areas" are as follows:

- Facilitation of traffic circulation
- Reduction of environmental pollution
- Small distance of parking spaces from driver's residence
- Extremely low cost of manufacture compared with other solutions
- Source of income for municipalities and a solution for parking problems
- Very profitable financial investment because of very short time of amortization of invested capital
- Exceptionally increased safety for the car provided by storage in underground mechanical parking
- Confrontation of stresses that accompany the citizen in his or her daily search for parking place (home, work, amusement, etc.)
- Possibility of release of 70% of street's surface from cars and output of this space to the municipality for exploitation

- Minimization of disturbance of street residents and shops owners because of short time required for manufacture
- Doubling or tripling of parking places for same street length depending on which type of system is manufactured

This invention also offers an alternate parking station system with a parking meter in places where many people gather, that is, courts, malls, and public services buildings, that are located in the center of the city. It also requires minimum financial investment because buying or expropriation of private building plots is not needed because it is constructed in a public area. It will also derive a substantial profit from selling or renting the parking areas or alternate parking with a time-limit parking meter. Finally, it's not causing a traffic disturbance on a large scale because the opening of the trench is done very quickly with modern excavating machines. After excavation, a waterproof and compact longitudinal tunnel is constructed by the molding, construction of supporting walls, the tunnel flooring, and roof. After completion, the street is immediately open to traffic. The remaining mechanical and electrical installations are completed underground inside the tunnel without any further disturbance to traffic. It is indicated that the surface parking spaces on the street remain as they were before.

Apart from the two systems discussed here, other systems are also available and supplied as per requirements.

7.4 EVALUATION CRITERIA OF A SITE FOR UNDERGROUND PARKING

The international tunneling association (ITA) in 1995 identified more than 20 sites where there was sufficient space for a car park. These sites were evaluated on the basis of the following criteria.

- *Accessibility* (Is the site on a main car traffic route? Will vehicles driving in and out cause congestion? Is public transport to important points in the city center available?)
- *Parking situation* (Is there high demand for parking at the site or in the area directly surrounding it, showing that a car park is needed? Has parking control already been introduced?)
- *Function for visitors* (How many functions that attract the public and how much shopping space are located within range of a car park on that site? For the moment, this range is assumed to extend out to a maximum walking distance of 500 m.)
- *Function for residents* (How many residents live within 500 m walking distance of the site?)
- *Relationship of the site to existing carports* (To what extent do the ranges of existing and already planned public car parks overlap with the range of the site under consideration?)

- *Relationship with intensification of land use* (Can an increase in the need for parking spaces in the immediate surrounding area be anticipated due to the construction or redevelopment of buildings?)
- *Zoning plan* (Do the conditions set forth in the zoning plan permit construction of a car park?)

Based on these criteria, which were also weighted to some extent (so that, for example, ease of access and the site's potential for serving visitors were weighted somewhat more heavily), a list of preferred sites can be drawn. Subsequently, feasibility studies should be carried out for preferred sites for final site selection.

The Environment Pollution (Prevention & Control) Authority (EPCA) in India has stated that underground parking facilities can be created under open spaces without disturbing the green areas on the surface and surrounding environment based on the site availability and after getting approval from the concerned agencies [4].

7.5 DESIGN OF UNDERGROUND PARKING FACILITIES

The layout design of a manually driven underground parking facility depends on the topography and the width and the length of the site, as well as the size of the premises. Examples of the following different types of architectural design of parking lots are found [5]:

- Simple, straight entrance and exit ramps
- Unending ramps
- Spiral entrance and exit ramps
- Modified versions of the aforementioned types

The *recommended slope* for straight ramps vary by situation such as (http://www.parkingconsultantsltd.com/rampfaqs.htm):

- Half-deck car parks where the vertical separation between decks is less than 1.5 m, slope is 1:6. This relatively steep slope is only possible when using transition gradients at the top and bottom.
- Where vertical differences are greater than 1.5 m, slope is not less than 1:10.
- Where ramps are curved, slopes are 1:10 or 1:12 depending on the separation.

In a manually driven car park, where people are likely to walk on ramps, any gradient steeper than 1:10 is likely to be problematic. Further,

- A person wearing shoes with elevated heels finds steep slopes very uncomfortable and possibly dangerous.
- People pushing shopping trolleys, buggies, or even bicycles find steep ramps very uncomfortable and, in many cases, dangerous.
- People with mobility challenges requiring aids such as walking sticks, crutches, or wheelchairs experience severe difficulties on ramps steeper than 1:12.

Many modern cars have wheel bases that are too long and underbody clearances of less than 50 cm. The effect of these specifications is that in any situation where a ramp of gradient steeper than 1:10 intersects with a flat slab, the bottom of cars will touch the transition line at the top of the ramp or vice versa (Fig. 7.8). Engineers have developed three-stage ramp structures to address some of these issues as follows:

- The top and the bottom of the ramp are constructed to a gentle gradient, say 1:16 or 1:20.
- The central section of the ramp is built to a steeper slope—1:8 or 1:10.
- Horizontal at the two ends.

Three additional issues with car park ramps are

1. Clear edge-to-edge width
2. The turning circle on approach routes to the bottom or from the top of ramps
3. The location of ticket machines on ramps

Many drivers find ramps too narrow and scrape their bumpers along walls at the top or bottom of ramps. The recommended minimum width for a one-way ramp is 3.0 m with an additional 0.3 m for side clearance to the structure. The recommended width of the entry section for a turning approach to a ramp is 3.5 m. Keeping in mind that although very few cars are more than 1.8 m wide, these recommendations allow for a broad range of driver behavior and skills.

The ITA [5] recommended that attention shall be given to user-friendly design, as well as security, for such parking facilities. Many people are uncomfortable in parking lots, particularly underground parking lots, because they are claustrophobic, afraid of geting lost, or afraid of mugging. Features that tend to enhance these fears are to be dealt with specifically and eliminated in the design of new parking lots.

For example, pillars, which impair vision and maneuvering, are to be avoided, particularly in parking facilities with a high turnover. An architectural design that eliminates pillars and extends over two rows of parking stalls and a traffic aisle creates the illusion of space and a greater sense of openness in the parking area.

(a)   (b)

FIGURE 7.8   The bottom of the car touches the ramp (a) at the end of the up ramp (b) at the start of the down ramp.

Because vacancies are visible from a farther distance, even sloppy parking results in only a minor loss of parking space and drivers' time.

With regard to the arrangement of individual parking stalls, one-way traffic, angled parking on both sides, a stall width of 2.30 m, and an aisle width of 5 m have proven to be user-friendly measures [5], despite the overall large spanning capacity (this construction mode results in a total structure width of 16 m or more). A diamond-shaped frame construction allows for low story heights of 2.70 m, as well as for relatively thin structures of 25 cm in the aisle area. In recent years, underground parking lots designated for short-term parking have been designed specifically as diamond-shaped frame structures. The construction costs have been reasonable.

Residential parking areas, reserved exclusively for long-term parking with little turnover, typically are executed as a modified, more economical version of the aforementioned arrangement, with stall widths of 7.50 m and individual supports for parking stalls. Compared to short-term parking facilities, a lower standard applies to residential parking lots.

To increase user-friendliness, the use of color and adequate lighting of parking decks are indispensable. Clear directional signs facilitate orientation to parking stalls. Wide stairways and elevators improve accessibility. Emergency calling systems, parking stalls reserved for women, and the presence of parking attendants make both underground and aboveground parking facilities more attractive and more acceptable to the general public.

7.6 PARKING GUIDANCE SYSTEM

In addition to structural and operational criteria, other factors that influence parking habits are the distance from a parking facility to the closest pedestrian zone and the means of charging and paying for parking. It has been observed in Stuttgart, Germany, that during peak times, many parking facilities reach their level of capacity, whereas others in the immediate vicinity still have some vacancies. In this connection, efforts have been aimed at improving this situation with the help of a parking guidance system designed to improve and better coordinate distribution of the vehicles among all parking facilities in the area. One goal of this plan is to improve the efficiency of those parking facilities located at the periphery and those that have a higher rate of vacancies.

The parking guidance system directs the motorists in search of a parking facility to the nearest vacancy, recommending the optimum route. A reduction in traffic searching for parking spaces helps reduce the impediments for through traffic, as well as reducing noise pollution and automobile exhaust fumes.

In 1988, all public parking lots in the Stuttgart city center became part of a dynamic parking guidance system, which was implemented in three stages. On the approach roads leading to the city center ring road, at the junctions, and the entrance roads to the parking facilities, variable message signs were mounted to provide motorists with up-to-date numeric information on vacancies in a given area or parking facility. The signs display two or three options in order to allow drivers to select the vacancy that is closest to their destination. Currently, a total of about 8000 parking stalls distributed throughout 22 underground and above-ground parking lots, as well as car parks, are connected to the parking guidance system. The completed system costs approximately 3.5 million DM, and annual operational costs are calculated at 150,000 DM.

Because of the modular structure of the guidance system, new facilities can be hooked up to it at any time. Data transfer stations installed at the control center of each parking facility enable operators to respond to changes in vacancies as they occur. Parking stalls reserved for permanent parking can be converted to short-term parking. The computer calculates the vacancies in all parking facilities every 2 minutes and then replaces the updated numeric display on every display field of the 45 variable message signs.

Traffic studies were carried out before and after the parking guidance system was installed. Results indicate that since the system was installed, the distribution of vehicles among all parking facilities has become more even. Waiting periods in front of the facilities have been reduced dramatically, and adverse effects on the environment caused by running cars have been halved. However, a survey of users of parking facilities revealed that while 85% of them are familiar with the parking guidance system, 35% do not take advantage of it.

7.7 PARKING LOT SECURITY

The following considerations may help designers implement sound parking measures for buildings that may be at high risk.

- Locate vehicle parking and service areas away from high-risk buildings to minimize blast effects from potential vehicle bombs.
- Restrict parking from the interior of a group of buildings.
- If possible, locate visitor or general parking near, but not on, the site itself.
- Restrict unauthorized personnel from parking within the secure perimeter of an asset.
- Locate general parking in areas that present the least security risks to personnel.
- If possible, design the parking lot with one-way circulation to facilitate monitoring for potential aggressors.
- Locate parking within view of occupied buildings.
- Prohibit parking within the stand-off zone.
- When establishing parking areas, provide emergency communication systems (e.g., intercom, telephones) at readily identified, well-lighted, closed circuit television-monitored locations to permit direct contact with security personnel.

- Provide parking lots with closed circuit television cameras connected to the security system and adequate lighting capable of displaying and videotaping lot activity.
- Request permits to restrict parking in curb lanes in densely populated areas to company-owned vehicles or key employee vehicles.
- Provide an appropriate setback from parking on adjacent properties if possible. Structural hardening may be required if the setback is insufficient. In new designs it may be possible to adjust the location of the building on the site to provide an adequate setback from adjacent properties.
- If possible, prohibit parking beneath or within a building.
- If parking beneath a building is unavoidable, limit access to parking areas and ensure that they are secure, well-lighted, and free of places of concealment.
- Do not permit uninspected vehicles to park under a building or within the exclusive zone. If parking within the building is required, the following restrictions may be applied:
  - Public parking with ID check
  - Company vehicles and employees of the building only
  - Selected company employees only or those requiring security

Operation of parking also varies by the charging method, that is, whether parking is free or charged; and, if charged, whether the fee is paid to a person or to a payment machine. Each case presents different risks, and the preferred security management necessarily differs depending on those risks. In general, self-driving parking facilities offer free access to users and to other people. Thus, car robbery, theft of belongings in a car, arson, and violent crime may have a high probability of occurrence in this type of parking lot. Also, accidents involving two or more cars or a car and a pedestrian are likely to occur. Access control of users, surveillance by CCTV, or patrol by personnel is necessary [1].

7.8 VENTILATION IN UNDERGROUND CAR PARKS

Enclosed (manually driven) car parks below natural ground level under large buildings present special ventilation requirements because generally natural ventilation cannot be utilized. Thus, mechanical ventilation is required to dilute or remove exhaust pollutants generated by vehicles with engines operating within the car park. Principal engine exhaust pollutants are carbon monoxide, hydrocarbons, and oxides of nitrogen. The dilution principle is used in car park ventilation to maintain a safe level of carbon monoxide within the car park space. In calculating the ventilation requirement for car parks, design standards are based on ventilation space volume, parking area, number of vehicles with engine operating, engine exhaust emission rate, and allowable carbon monoxide exposure level and duration. In multilevel underground car parks where the number of cars per level may exceed 500, the quantity of ventilation air required may be very large. It is understood that sensible design may be obtained with a balance between energy conservation and allowable carbon monoxide exposure level and duration.

Studies of air quality usually have concentrated solely on carbon monoxide because it is generally considered to be the major health factor involved. However, other pollutants, such as oxides of nitrogen, oxides of sulfur, and odorous compounds, also contribute to poor air quality. In the United States, basic design criteria, as quoted by [6], are 4 to 6 air changes per hour, with a minimum ventilation rate of 7.5 liter/s/m^2, and a maximum ventilation rate of 10 to 15 liter/s/m^2. The maximum recommended carbon monoxide concentration is 50 ppm for an 8-hr, time-weighted average and 125 ppm for a 1-hr, time-weighted average. In Japan, as quoted by [7], the mean carbon monoxide level per 8-hr average is 20 ppm; the environmental standard for inhabitable rooms is 10 ppm over the same period. The ventilation requirement for car parks in Finland is determined by the carbon monoxide concentration [8]. The limit for carbon monoxide concentration is 30 ppm for an 8-hr average and 75 ppm for a 15-min average. The minimum supply air flow rate is 2.7 liter/s/m^2. This rate would require 9.72 air changes per hour per meter length of ceiling height. Australian Standard [9] specifies that a minimum flow rate of fresh air of 3000 liter/s is to be maintained. This would give an indoor carbon monoxide level of about 136 ppm if a loading value of 52 ft^3/hr (0.4 liter/s or 1.4 m^3/hr) of carbon monoxide discharged, as proposed by [10], is used. Australian standards also specify the average driving lengths of cars for a heavy-loads car park.

For these large enclosed spaces, only a small portion of the volume space would be ventilated, except at the occupant levels. Therefore, conventional macroscopic numbers, such as the number of air changes per hour per floor area, are not sufficiently accurate in specifying the ventilation requirement of the area. Also, increasing the ventilation rate alone might not be effective in reducing the contaminant level in a given facility [11].

The concentration levels of carbon monoxide (CO), nitrogen dioxide (NO$_2$), and particulates (PM10 and PM2.5) in underground and elevated car parks in Kota Kinabalu city were studied for a period of 6 months [12]. The maximum 15-min average concentration of CO in underground car parks was found to be higher than in elevated car parks, both during weekends and weekdays. Meanwhile for NO$_2$, maximum concentrations were comparatively higher in elevated car parks than in underground car parks. Results have shown that none of the CO and NO$_2$ concentration levels exceeded the Swedish STEL (15-min average) of 100 and 5 ppm, respectively. However, based on an 8-hr average, only the CO concentration (26.0 to 29.5 ppm) in the WP underground car park during weekends has exceeded the Swedish hygienic limit (20 ppm). Most of the time, maximum concentrations of PM10 and PM2.5 in the WP underground car park during weekends and weekdays were observed to be higher than in the CP elevated car park. Generally, the highest concentration levels of all pollutants were observed to occur during rush hours. Concentrations of CO and particulates

TABLE 7.2 Recommended Indoor Environment [14]

Condition	Recommendation
Temperature	19–23°C
Relative humidity	40–70%
Air speed	0.1–0.3 m/s
Ventilation rate	8 liter/s

(PM10 and PM2.5) in both car parks have indicated good correlations with the number of cars entering and leaving the car parks. These results have suggested that a relationship exists between pollutants and the number of cars. Meanwhile, weak correlations between NO_2 concentrations with the number of cars in both car parks were found, which could suggest that other factors have influenced the pollutant concentrations.

Chow [13] conducted a study on the indoor environment of 19 enclosed car parks. The three important indoor environmental parameters, namely mean air temperature, mean relative humidity, and mean air speeds at the occupied zone, were measured to obtain a general picture of the indoor environment of the car parks. Car park users were surveyed to give a four-point assessment—classified as Very Comfortable (VC), Comfortable (C), Acceptable (A), and Uncomfortable (UC)—on their subjective feelings concerning the thermal environment within the car park facilities.

Results of the study indicated that car parks having a combination of a mean air temperature greater than 26°C, mean relative humidity higher than 65%, and mean air speed lower than 0.4 m/s would be uncomfortable for people using car parks in Hong Kong. Air speed is more than the range given by [14] in Table 7.2.

7.9 ECONOMICS OF UNDERGROUND PARKING FACILITIES

Figure 7.9 shows the average construction cost per vehicle, including land acquisition cost, for certain ranges of parking capacity for the three locational types of parking facilities. Because these are average costs, based on data for 319 facilities throughout Japan and computed without separating out mechanized facilities, which include expensive equipment, results are extremely rough. However, the following characteristics can be deduced:

- Construction costs for small capacity underground facilities are extremely high.
- For surface and aboveground parking facilities, the land acquisition portion of total construction cost is of a magnitude that cannot be ignored. However, the figure indicates that the land acquisition cost is zero for surface facilities

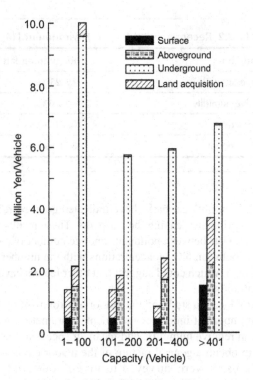

FIGURE 7.9 Construction cost of parking facilities in Japan [2].

that have a capacity of 410 or more vehicles. In fact, this finding represents only a single facility, which was constructed on a former river bed, that is, on existing public land.

- The construction cost for facilities of large capacity is high, irrespective of their locational type.

The fact that construction costs for underground parking facilities are very high may reflect the unique ground conditions that exist in Japan. Almost all large cities in Japan are located on deep quaternary deposits that are low in bearing capacity, thus necessitating significant construction expenditures for ground support and water cutoff at the excavated surface.

7.10 CASE HISTORIES

7.10.1 Sydney Opera House Underground Car Park

The Sydney Opera House is an important landmark in the historic area of Sydney. Its location on Bennalong Point commands an impressive view of the Sydney Harbour Bridge, and it adjoins Government House grounds and the Royal Botanic Gardens. These features, in addition to its exciting architecture and renowned cultural venue, attract numerous visitors: every year, nearly 1.4 million patrons and 2 million tourists visit the opera house.

Unfortunately, features that add to the appeal of the opera house have also created significant difficulties in providing vehicle access to it. In particular, the lack of car parking facilities presented a problem after the opera house opened in 1973.

These parking needs are now met by the Sydney Opera House car park, known as the Bennelong Point Parking Station (Fig. 7.10). This car park is the first helical underground parking station, comprising 12 stories and providing 1100 parking spaces (Fig. 7.11). It consists of a free-standing, double-helix

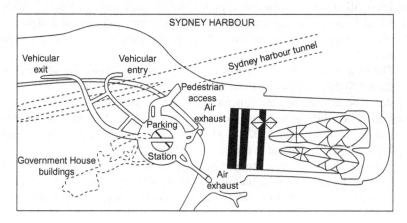

FIGURE 7.10 Location plan of the Sydney Opera House underground car park [5, 15].

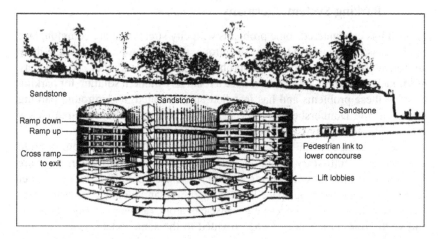

FIGURE 7.11 Architect's sketch of the double-helix Sydney Opera House underground car park. *(source: earthmover and civil contractor; project report in [5])*

concrete structure, wrapped around a central intact core containing linking drives and service tunnels. The huge, doughnut-shaped cavern has been created with an outer diameter of 71.2 m and an inner diameter of 36.4 m. The structure is 32 m high and extends some 28 m below sea level.

Vehicle access to the parking station is via tunnels from Macquarie Street, which had to pass over the top of the Sydney Harbour Tunnel. This constraint dictated that the cavern be constructed as close as possible to the ground surface.

In order to avoid detracting from the aesthetic appeal of the area, all work had to be done without disrupting the surface, within either the Botanic Garden or the forecourt of the opera house. Because the cut-and-cover method was precluded by these environmental constraints, the car park was constructed within the Royal Botanic Gardens, entirely underground. One feature of the design was that the existing established fig trees above the car park should not be affected by its construction or operation.

The rock cover above the cavern is just 7–9 m thick, consisting of variably weathered Sydney sandstone, overlain by 1 to 2 m of soil of the Royal Botanic Garden. The helical concrete structure does not provide long-term support for rock cover; the largest unsupported span of the same is 17.4 m. The various challenges involved in the design and construction of this car park have been reported at conferences and in technical literature [15].

This world-famous case study significantly illuminates the advantages of underground structures in achieving solutions that simultaneously preserve the environment and meet the parking needs associated with much visited public places.

7.10.2 Munich Automated Underground Parking System, Germany

Munich has encountered some problems with city streets that are commonplace right across Germany and the rest of Europe, that is, historic cities where there are narrow streets and a congestion problem due to the parking of cars. The city of Munich decided in 2004 to introduce a relatively modern solution for parking to solve these problems and has installed an automated, underground parking system in the Donnersbergerstraße in the München–Neuhausen area (www .roadtraffic-technology.com/projects/munich/).

The advantages of this can be seen immediately: there is no room to build conventional multi-story car parks aboveground in the areas where they are required and understreet parking is actually more cost effective than other schemes on a per parking space basis. In addition, an automatic system brings significant savings in engineering because elevators, ramps, stairways, lighting, and ventilation do not need to be provided to the same extent. Construction, however, also means the transfer of sewer, storm drains, and water services from beneath the center of the road to conduits at the side of the roads.

Design of the facility started in January 2004 with construction beginning in October 2004. Construction was completed by February 2006 (road works took 10 months and the entire project took 16 months from the start of construction). The investment for the project was €11.35 m (45% on building, 30% on the parking system, and 25% on other costs) (the investor was Landeshauptstadt München).

The parking system provided is a combination of two Wöhr Multipark 740 systems, which provides 284 parking places (150 plus 134). The equipment was supplied by Wöhr, and installation was carried out by the general contractor/engineer Wöhr and Bauer GmbH.

The parking system will be run and maintained by Park and Ride (P and R) GmbH for a contracted period of 20 years. The consultant for the project was GiVT Gesellschaft für innovative Verkehrstechnologie mbH of Berlin; the project structural engineer was ISP Scholz Beratende Ingenieure AG of Munich.

Dimensions of the underground building are 121 m in length, 15 m in width, and 10 m in depth. There are 284 parking bays on four parking levels with two transfer stations and four elevators and shelf control devices. Retrieval time for vehicles is on average 120 seconds for the whole process. The car park was excavated and constructed using cut-and-cover methods along with precast/prestressed reinforced concrete wall sections, roof beams, and cover sections.

Parking Process

Cars (maximum dimension 5.25 m long, 2.2 m wide, and 1.7 m high) must be driven to one of four receiving stations (marked A to D), the entry to which is actuated by a control device that responds to a transponder chip held in the car by the user.

On gaining entry, the driver proceeds to correctly position/park the car on a transfer ramp and then locks and leaves the vehicle. A combination of laser scanners and light barriers will then examine the car for its positioning and dimensions. An elevator ramp is then actuated and the car is raised to its parking level and stored (cars are parked side by side).

On returning, the driver comes to the same transfer station (he or she may pay by credit card at an automated paying station) and the car is retrieved from its storage level according to the transponder chip still held by the driver. The car is retrieved on its transfer pallet and the driver simply drives away through automated exit gates. The pallet has a lighting system that illuminates the area for 2 minutes while the driver gets into the car.

Customer help is available via an intercom system 24 hours per day in case of breakdowns or unforeseen problems. If a car is not retrieved for its owner within 2 hours then the owner receives €100 compensation from the parking operator.

Underground parking facilities should also be an integral part of an underground metro system. Most mega cities of the world are integrating facilities of underground parking with metro and other road tunnels. The next chapter is devoted to underground metros and road tunnels.

REFERENCES

[1] Ishioka H. Security management for underground space. Tunnel Undergr Space Technol 1992;7(4):335–8.

[2] Fukuchi G. At or below ground level? An example of urban parking facilities in Japan. Tunnel Undergr Space Technol 1994;9(1):53–57.

[3] Ikaheimonen P, Leinonen J, Marjosalmi J, Paavola P, Saari K, Salonen A, Savolainen I. Underground storage facilities in Finland. Tunnel Undergr Space Technol 1989;4(1):11–15.

[4] EPCA. Implementation of the parking policy, environment pollution (prevention & control) authority (EPCA) report No. 25 (July 2006). In response to the Hon'ble Court order dated May 5, 2006.

[5] ITA. Underground car parks: international case studies, report of ITA working group No. 13 on direct and indirect advantages of underground structures. Tunnel Undergr Space Technol 1995;10(3):321–42.

[6] Stankunas AR, Bartlett PT, Tower KC. Contaminant level control in parking garages, American Society of Heating, Refrigeration, and Air-Conditioning Engineers paper DV-80-5, No. 3. RP-233, May, 1989, 584–605.

[7] Matsushita K, Miura S, Ojima T. An environmental study of underground parking lot developments in Japan. Tunnel Undergr Space Technology 1993;8:65–73.

[8] Koskela HK, Rolin IE, Norell LO. Comparison between forced displacement and mixing ventilation in a garage. American Society of Heating, Refrigeration, and Air-Conditioning Engineers 1991; part 1A:1119–26.

[9] Australian Standard. Ventilation of enclosures used by vehicles with internal combustion engines, Standards Australia (Standards Association of Australia) 1 The Crescent, Homebush, NSW 2140, No. 1668.2, section 4; 1991.

[10] Ball D, Campbell J. Lighting, heating and ventilation in multi-storey and car parks. In: Proceedings of the joint conference in multi-storey and underground car parks. London: Institution of Structural Engineers and the Institution of Highway Engineers; 1973. p. 35–41.

[11] Chow WK. On ventilation design for underground car parks. Tunnel Undergr Space Technol 1995;10(2):225–45.

[12] Sentian J, Ngoh LB. Carbon monoxide (CO), nitrogen dioxide (NO2) and particulates (PM10 and PM2.5) levels in underground and elevated car parks in Kota Kinabalu City. In: Brebbia CA, editor. Air pollution XII. WIT Press, www.witpress.com; 2004.

[13] Chow WK, Fung WY. Survey on the indoor environment of enclosed car parks in Hong Kong. Tunnel Undergr Space Technol 1995;10(2):247–55.

[14] Chartered Institution of Building Services Engineers. Environmental criteria for design, CIBSE guide, section A1, London; 1987.

[15] Pells PJN. The Sydney Opera House underground parking station. Int Soc Rock Mech (ISRM) News J 1993;1:2.

Chapter 8

Underground Metro and Road Tunnels

It is cheaper to do things right the first time.

Phil Crosby

8.1 INTRODUCTION

A new era of underground space technology has begun with extensive networks of underground metro systems all over the world. London's metropolitan railway was the world's first subway. The 6-km section was opened in 1863 and ran between Paddington and Farrington. No mega city may function efficiently without a mass transit system of high performance. They offer to everyone fast, safe, comfortable, and inexpensive access to different areas of a city. The modern underground metro systems are more beautiful, brightly lighted, well developed architecturally, and hygienic, going right to the heart of the mega cities, and quite safe. Undercity tunnels are also being excavated for direct bypass traffic as in Australia.

Following are the advantages of underground metros in the mega cities [1]:

- Crossing of hills, rivers, and a part of oceans (straits).
- Increase in market value of adjacent land and savings in man hours.
- They also favor a more aesthetic integration into a city without blocking views of beautiful buildings, bridges, monuments, and religious functions.
- Very high capacity in peak hours in any direction. They form a part of an integrated total city transportation system for the convenience of people. They may reduce traffic jams on the surface. Metro service, therefore, should be encouraged and extended to airports.
- They protect residents completely from severe round-the-clock noise pollution from surface traffic.
- Efficient, safe, more reliable, faster, comfortable, tension free, environmentally sustainable, and technically feasible in developing nations. They require just 20% of the energy that is consumed by road traffic. They reduce road accidents and pollution due to a decrease in vehicular traffic.
- Underground metro and road tunnels help in emergency services of medical doctors to patients, etc.

Regularity, punctuality, and speed are basic criteria in determining the attractiveness of public transport. Success is very apparent where route separation has permitted a demixing of public and private traffic. This has cut travel times compared with bus and tram services, reduced delays to zero in most cases, and increased passenger loads (Table 8.1).

The objectives of mass transit systems, with examples of associated "impacts and requirements," can be summarized as given in Table 8.2.

With more and more use of underground transit systems, it is necessary to prepare contingency plans accordingly to take care of emergency situations. A good example of this is the blackout that occurred in the United States and Canada on August 14, 2003. More than a thousand persons were stranded in subways, but police had been trained to evacuate people from subways and skyscrapers without an increase in panic.

Opinion polls carried out in the United States, Japan, and 14 European countries showed clearly the public moral support for environmental protection, even at the expense of reducing economic growth. The fatigue strength of concrete should be considered in the design of superstructures and bridges. An elevated metro corridor destroys the ecosystem, aesthetics, and general landscapes of the areas and increases noise pollution. Hence public opinion is in favor of underground metros. The underground metro of Beijing may be bigger than that of London and New York by 2015.

The metro rail system uses ballastless track without joints, which makes it almost free of maintenance. Signals are in the driver's cabin only. Because the

TABLE 8.1 Transport Improvement due to Construction of Urban Railways in Germany [2]

Transport Form	Measure	Result
Travel time		
Underground/light railway	Bus line replaced by rail system	Average travel time cut by one-half
Delays		
Underground/light railway	Tunnel or separate right of way	Delays of up to 10 min cut to virtually none
Passenger loads		
Underground/light railway	Tunnel or separate right of way	Rise in passenger loads Normal: +15 to 16% Extreme: Frankfurt (+109%), Munich (+190%)
Commuter railway system (S-Bahn)	Separate right of way	Rise in passenger load +100 to 250%

TABLE 8.2 Objective, Mass Transit Impact, and Mass Transit Requirements

Objective	Mass Transit Impact	Mass Transit Requirement
Socioeconomic efficiency	1. Passenger time savings	Must attract bus, auto, or other mode passenger; required to be rapid, relatively inexpensive, and reliable
	2. Less traffic congestion	
	3. Cost saving to society	
	4. Facilitates commerce and growth of city	
	5. Increases productivity	
Support for city development plan	6. Allows urban areas to function more effectively	High passenger capacity per hour; increases transportation capacity between city centers and urban regions
	7. Can direct or influence urban development	
Social impacts and improvements	8. Land acquisition and relocation during construction	Accessibility for most. This requires good pedestrian access (ramps/lifts). Design for public safety must be satisfied
	9. Provides access for all (including elderly and disabled people, etc.)	
Environmental improvement	10. Depends on urban character, quality of metro design, environmental laws, and regulations	Identification of an alignment that contributes to the planned, holistic improvement of a corridor
	11. Underground allows pedestrianization and surface enhancement	Use of construction methods, which mitigates adverse construction impacts
	12. Reduction of surface impacts, e.g., noise, pollution, visual	Good design quality

software controls automatic driving of an engine, trains stop at exactly the same position within ±10 cm. As a result, disabled persons may enter the coaches comfortably. The driver only opens and closes doors. Underground stations (with cross passage below tracks) are air-conditioned. There are parks above underground stations. Performance is the best publicity. The life of a metro is about 100 years. Underground stations should meet fire safety and evacuation norms [3].

Tunnels are ventilated properly. One fan pumps air and another fan on the other side acts as exhaust to take out smoke in case coaches catch fire. These fans are switched on by station masters. The train will move to the pumping fan side so that passengers do not die of smoke inhalation. All coaches are connected with see-through ends for further escape.

Unfortunately, construction costs for underground systems are a major deterrent when city officials consider the option of an underground metro. Table 8.3 compares relative costs of the various types of infrastructure on the basis of a study conducted by the French Tunnelling Association.

It is the experience of road users that the open cut method of construction leads to a lot of the inconvenience to a society and disruption of the environment, which must be compensated financially if any justifiable comparison between the cut-and-cover method of construction and tunnel boring is to be made.

Figure 8.1 shows a rail metro tunnel. Prefabricated lining shown in Figure 8.1 is suitable for various soil, boulder, and rock conditions, except squeezing

TABLE 8.3 Relative Costs of Interstation Structures

Location	Infrastructure (a)	Equipment	Total (b)	Ratio (a)/(b)
At grade (surface)	25	30	55	0.45
Elevated (super-structure)	100	30	130	0.75
Long span bridge	250	30	280	0.90
Cut and cover	100 to 200	40	140 to 240	0.70 to 0.80
Tunneled	150 to 500	50	200 to 550	0.75 to 0.90

FIGURE 8.1 Precast lining in a metro tunnel. *(source: http://www.railwayage.com/sept01/washmetro.html)*

grounds (due to the high overburden pressure) and flowing grounds within water-charged, wide-shear zones (due to seepage erosion or piping failure). These may not occur in shallow tunnels.

The work culture of the Delhi Metro Rail Corporation is that there is no clerk with very few peons, which is the key to success. There is no witch hunting for a wrong decision. The decisions are not delayed. Wrong decisions are noticed and get corrected automatically. Punctuality of staff plays an important role. NATM was adopted. Ten trees are planted for every cut tree. All underground stations are built by the cut-and-cover method. The entire site is closed by walls on all sides. Exhaustive instrumentation is done to learn lessons for the construction of future metros. The rehabilitation of structures (damaged by subsidence along tunnels) is the responsibility of all contractors to save excessive time that is lost in litigation by management (see http://www.delhimetrorail .com). Five metros of Delhi, Hong Kong, Taipei, Singapore, and Tokyo make an operating profit out of 135 metros all over the world. The price of land rises near a metro.

8.1.1 Findings of International Tunnelling Association

The International Tunnelling Association [4] has presented the following observations after analysis of data from 30 cities in 19 countries.

- The typical cost ratio for surface, elevated, and underground metro systems was found to be approximately 1:2:4.5.
- It is generally accepted that underground systems are more expensive to operate than elevated or surface systems.
- Because of the large investment required (capital and recurring costs) and significant urban and environmental impacts, the choice is nearly always resolved politically. The government has to subsidize the cost to reduce the cost of a ticket. The metro is not viable commercially.
- The overwhelming choice (of 78% alignment) for urban metro systems is underground with very little at grade (surface) alignment. They are typically designed to be of high speed and capacity (20,000 passengers per hour per direction) serving the city center.
- In many cases—for example in the center areas of older cities (with 2 to 7% area of streets only)—for functional, social, historic, environmental, and economic reasons, there is no alternative to the choice of an underground alignment for new transit systems.
- Noiseless technology may be used in the tunneling.

In the 21st century, new economic centers are being developed by the global finance corporations in adjoining small towns, thus avoiding the hassle of long traffic jams, high wages of skilled and nonskilled workers, high cost of buildings and infrastructure and safety, etc. in the developing nations. The small inexpensive Nano car is revolutionizing the traffic system in India.

8.2 TUNNEL BORING MACHINE (TBM)

After nearly 150 years of development, the TBM has been perfected to excavate in fair to hard rock masses. Tunnel boring machines with features purpose built to the specific ground conditions are now the preferred mode for bored tunneling in mega cities. The high capital cost is justified when tunnel length exceeds 2 km [1]. These TBMs offer the following advantages over the drilling and blasting method in metro tunnels.

- Explosives are not used, hence operations in densely built-up areas produce much lower vibrations.
- Excavation is fast. Time is money.
- Lower initial support capacity saves cost.
- Less labor cost.
- Reduces surface settlement to very low levels, resulting in assured safety to the existing super structures.
- Reduces risk to life of workers by (i) rock falls at face or behind the TBM, (ii) explosives, (iii) hit by vehicles, and (iv) electrocution.
- Reduction in overbreaks.
- Minimum surface and ground disturbance.
- Reduced ground vibrations cause no damage to nearby structures, an important consideration for construction of an underground metro.
- Rate of tunneling is several times that of drill and blast method.
- Better environmental conditions—low noise, low gas emissions, etc.

Engineers should not use TBM where engineering geological investigations have not been done in detail and the rock mass conditions are very heterogeneous. The contractor can design the TBM according to the given rock mass conditions, which are normally homogeneous nonsqueezing ground conditions. The TBM is unsuitable for squeezing or flowing grounds [5]. There are reports of a specially designed shielded TBM for squeezing ground conditions.

8.2.1 Tunnel Boring Machines for Hard Rocks

The principle of the TBM is to push cutters against the tunnel face and then rotate the cutters for breaking the rocks in chips.

The performance of a TBM depends on its capacity to create the largest size of chips of rocks with the least thrust. Thus rock chipping causes high rate of tunneling rather than grinding [6]. The rate of boring through hard weathered rock mass is found to be below expectation.

Disc cutters are used for tunneling through soft and medium hard rocks. Roller cutters are used in hard rocks, although their cost is high. A typical TBM is shown in Figure 8.2, together with ancillary equipment. The machine is gripped in place by legs with pads on rocks. Excavation is performed by a cutting head of welded steel and a convex shape, with cutters arranged on it

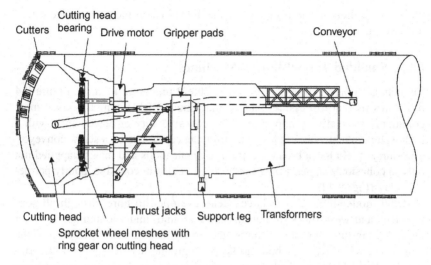

FIGURE 8.2　Tunnel boring machine and ancillary equipment [7].

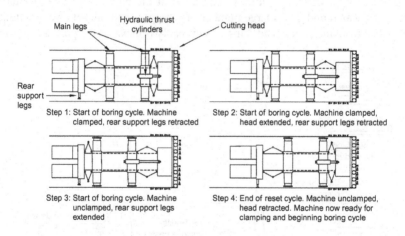

FIGURE 8.3　Method of advance of a rock tunnel boring machine [7].

optimally. The long body of the TBM contains four hydraulic jacks to push forward the cutting head and also drive motors that rotate the cutting head for chipping rocks. Figure 8.3 shows schematically a method of advance of the cutter head and shows how the TBM is steered and pushed ahead in self-explaining four steps. Typically, even when a TBM operates well, only 30 to 50% of the operating time is spent on boring.

Figure 8.2 also shows the removal system for muck (rock chips). The excavated material is collected and scooped upward by buckets around the cutter head. These buckets then drop the rock pieces on a conveyer belt, transporting them to the back end of the TBM. There they are loaded into a train of mucking cars.

8.2.2 Shielded Tunnel Boring Machines

Tunnels are bored through soils by the shield tunneling method. Figure 8.4 illustrates the earth pressure balance machine (EPBM), which is used most commonly in relatively impervious and cohesive silty or clayey soils with a high water content. Muck is removed from the cutter face by a screw conveyor. The slurry shield has a broad spectrum of applications and may be applied not only in cohesionless, permeable sandy soil, but also in cohesive, poorly permeable clay (Fig. 8.4d).

Dual-mode shield TBMs have been developed to bore through all soil, boulders, and weak rocks (in nonsqueezing ground) under a high groundwater table. During tunneling the groundwater table is lowered to the bottom of the tunnel by drilling drainage holes to keep the ground dry. It works on the principle of shield TBM on which scrapper picks as well as disc cutters are mounted on the cutter head. Table 8.4 summarizes the salient features of dual-mode TBM and EPBM. During initial excavation at the New Delhi underground metro, it was found that a large number of scrappers and buckets were getting detached from the cutter head. This was probably because of the presence of too

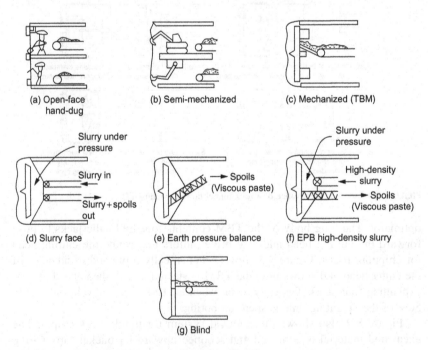

FIGURE 8.4   Types of shields in tunnels through soils [8].

TABLE 8.4 Salient Features of Tunnel Boring Machines [9]

Serial No.	Item	Earth Pressure Balance Machine (EPBM) (Fig. 8.4e)	Dual-Mode TBM— Shielded Tunnel Boring Machine
1	Manufacturer	Herrenknecht of Germany	Herrenknecht of Germany
2	Diameter	6.490 m	6.490 m
3	Length of shield	3.8 m (7 m including tail skin)	3.9 m (6.9 m including tail skin)
4	Weight of shield	252 MT	325 MT
5	Length of complete system	57 m	70 m
6	Cutter head rotation	1 to 7 rpm	1 to 6 rpm
7	Torque	4000 KNm	4377 KNm
8	Tunnel lining	Precast segmental RCC	Precast segmental RCC
9	Finished diameter	5.7 m	5.7 m
10	No. of segments per ring	6 (5+1 key)	6 (5+1 key)
11	Thickness of lining	280 mm	280 mm
12	Length of ring	1.2 m	1.2 m
13	Grade of concrete	M-45	M-45
14	Weight of each ring	16 tons	16 tons
15	Joint sealing	EPBM gasket and hydrophilic seal	EPBM gasket and hydrophilic seal
16	Power required	3 MW for each machine	3 MW for each machine
17	Planned progress	10 m per day	6 m per day
18.	Maximum progress achieved so far	28.8 m per day	7.2 m per day

many boulders in the soil strata. As a result, bigger boulders were entangled in the large space between the arms, thereby knocking off the scrapper and buckets. Then protective plates and deflector strips were added around the buckets to avoid a direct impact of boulders on the buckets, in addition to other modifications. Thereafter a dual TBM has succeeded [9]. The advantage of a fully shielded TBM with segment erector is that there is no unsupported ground behind the shield, which is why TBMs have failed in poor grounds yet dual

TBMs succeeded [10] in soils, boulders, and weak rock mass in a nonsqueezing ground condition (H < 350 $Q^{1/3}$ m, where Q is rock mass quality). Delhi Metro Rail Corporation (DMRC) purchased 16 TBMs from Herrenknecht of Germany for excavating a large number of metro tunnels in India. DMRC has suggested an eco tax on petrol to provide funds for underground metros all over India.

It is necessary to inject foam along with water at the cutter head, which has the following advantages:

- Reduced permeability and enhanced sealing at the tunnel face.
- Suppresses dust in the air in rock tunneling (due to high content of SiO_2 in rocks).
- Excavation of wet soil or weathered rock is easier.
- Soil does not stick to the cutters.

A double-shielded TBM has been developed by M/s Mitsubishi, Japan. A Japan boring shield can pass through the outer shield easily, if curves are easy in the tunnel alignment. It advances with concurrent concrete segmental lining. It has been used successfully for hydroelectric tunnels in Spain, Ecuador, Lesotho, and India in complex geological conditions and shallow cover [11]. It succeeded at the Parbati II Hydroelectric Project, although the open TBM failed there. Figure 8.5 also shows that a double-shielded TBM is able to bore through weak rocks faster than an open TBM [12]. The shielded TBM is an expensive machine but cost-effective for tunnels longer than 1.5 km with $q_c < 45$ MPa (where qc is uniaxial compressive strength of intact rock material). It may be, therefore, used in metro, rail, and road tunnels in highly variable complex geological conditions. Singh and Goel [13] have summarized the rock mass

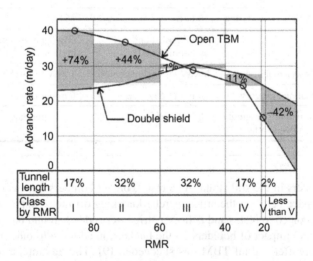

FIGURE 8.5 Variation of advance rate of open TBM and double-shielded TBM with RMR in Evinos tunnel [12].

excavability index developed by Professor Z.T. Bieniawski and Q_{TBM} by Dr. N. Barton for predicting approximately the rate of tunneling by a TBM.

There is no experience of success in TBM tunneling through squeezing grounds anywhere in the world. It is understood that a TBM may be stuck in highly squeezing ground or flowing ground. Therefore, both TBM and shielded TBM are not recommended in squeezing ground and flowing conditions.

Obviously water lines, sewer lines, and so on have to be protected during tunneling at shallow depths below congested mega cities. Sometimes a sewer line is ruptured during tunneling. Enormous stinking sewerage is spread on the roads. It is difficult to repair sewer lines quickly. In soil areas, the ground-water table (GWT) is lowered below the tunnel base before tunneling is done in relatively dry soil. The subsidence profile due to lowering of the GWT is, however, wider compared to that due to tunneling. Careful underpinning of the foundations or columns of the old cracked building is done to adjust to the subsidence increasing with time, such as at Delhi.

There is a tendency very often to term "geological surprises" as the cause to justify time and cost overruns in the completion of tunneling projects. This could be true in some cases, yet managers should be cautious.

The TBM is an extremely useful technological advancement. No doubt the technology advancement leads to the cultural fusion, and countries with different cultures are coming closer. One such example is the Channel tunnel. The 49.9-km-long tunnel between the United Kingdom and France was excavated through chalk marls from the United Kingdom side and through water bearing tectonically affected chalk from the France side by TBMs. Work on one of the largest private sector projects was started at the end of 1986. On December 1, 1990, the historic junction was made between France and Great Britain. On May 6, 1994, Queen Elizabeth and President Mitterrand inaugurated the Channel tunnel [14].

8.3 PRECAST LINING

In some projects, fiber-reinforced precast concrete linings have been adopted. Precast concrete segmental lining is now used in soil, boulders, and weak rock masses (Fig. 8.6). The TBM is capable of placing them in position all

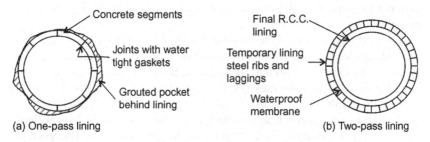

(a) One-pass lining   (b) Two-pass lining

FIGURE 8.6 Tunnel lining in soils [8].

around a circular tunnel with the help of a segment erector. Segment bolts are then tightened twice by impact wrenches. The curved alignment is achieved with the help of tapering of the lining rings. All rings are tapered, and curvature is obtained by adjusting the orientation of rings suitably. Before taking them inside the tunnel, the segments are checked on the ground for any cracks/ damage. As water tightness is extremely important for the durability of the tunnel lining, a double gasket system comprising a durable elastomeric gasket and a water sealing made from hydrophobic material is used. These gaskets are located in grooves cast into the edges of the precast concrete segments. Together with high precision casting of the segments achieved by precision steel moulds, gaskets will ensure durable and water tight tunnels. Hydrophobic seals expand up to 250% once they come into contact with water.

Thought should be given to a fire-resistant design of concrete lining, as fires in trains are common these days. An extra thickness of concrete covers ($\cong$ 75 mm) should be provided over the steel reinforcement. Underreinforced concrete segments may be used to ensure the failure in the ductile phase, if it occurs.

Grouting is carried out simultaneously with tunneling. There are inbuilt ports in the tail skin of the TBM. These are used in primary grouting of the annulus (void between excavation profile and outer face of the precast ring). Grouting is continued up to 3 bars (0.3 MPa) pressure. Excavation is not commenced until the previous lining is completed. Secondary grouting is also done within 14 days of ring erection. Every third ring is grouted to a pressure of 3 bars (0.3 MPa). Secondary grouting will fill up any void left during the primary grout due to its shrinkage.

8.4 BUILDING CONDITION SURVEY AND VIBRATION LIMIT

Open trenches and shafts are excavated by the drilling and blasting method for connection to the underground metro system. The controlled bench blasting method is used in open excavation, under busy and congested roads that are flanked by old or heavy buildings and monuments. Before designing the controlled blasting, the entire rock mass is explored thoroughly. Trial blasts are detonated to determine the safe-scaled distance (= $R/W^{1/2}$, where W is the weight of charge per delay of detonators and R is the distance from the blasting pattern) to control peak particle velocity according to the nature of the structures.

The next step is to assess the condition of buildings standing near the blast site to determine how much vibration can be sustained by these structures, especially old buildings and ancient monuments, if any. Table 8.5 specifies permitted peak particle velocities (PPV) as per German standards. It may be reminded that ISRM has recommended almost twice the PPV values.

Archaeologists suggest that no surface metro station should be built within a protected 100-m periphery of a protected (heritage) monument. However, an underground metro station may be a better choice there.

TABLE 8.5 Permitted Peak Particle Velocities (PPV) on Structures

Serial No.	Condition of Structure	Maximum PPV in mm/s
1	Most structures in "good condition"	25
2	Most structures in "fair condition"	12
3	Most structures in "poor condition"	5
4	Water supply structures	5
5	Heritage structures/bridge structures	5

8.5 IMPACT ON STRUCTURES

Blasting works may affect the surrounding structures slightly, despite controlled blasting. In a worst case, small cracks may develop in roller-compacted concrete and masonry. The air overpressure may also create cracks in the glass works of doors and windows in nearby areas. Table 8.5 summarizes the various types of damages to structures. Huge compensation may have to be given to owners of damaged nearby buildings according to the specified class of damage [15].

Traffic is stopped during blasting time for a few minutes only, and all roads and other exits/entries to the blasting site are closed for safety reasons. The flying of rock pieces during an urban blasting may have severe consequences.

8.6 SUBSIDENCE

The subsidence of ground and differential settlement of nearby structures takes place due to underground tunneling. Dewatering due to excavation causes more widespread subsidence due primarily to the settlement of overlying loose deposits of soil, silt, or clay. In totally rocky areas, the subsidence is very small and does not cause any worry. The following instruments are recommended for precision monitoring of structures.

- Precise leveling points
- Tiltmeters
- Crack gauges embedded in the nearby structures
- Vibration monitoring of old/ancient structures

Further blasting may also be the cause of subsidence of ground, especially loose soil slopes where differential settlement may damage surface superstructures.

In case actual settlement is expected to go beyond the predicted subsidence, the whole construction methodology must be reviewed. Table 8.6 may be used, which specifies the maximum tensile strain caused by subsidence (= increment in spacing of columns divided by the distance between columns, expressed as a percentage).

TABLE 8.6 Building Damage Classification [15]

Risk Category	Description of Degree of Damage	Description of Typical Damage and Likely Form of Repair for Typical Masonry Buildings	Approximate Crack Width (mm)	Max Tensile Strain % due to Subsidence
0	Negligible	Hairline cracks		Less than 0.05
1	Very slight	Fine cracks treated easily during normal redecorations. Perhaps isolated slight fracture in building. Cracks in exterior brickwork visible upon close inspection.	0.1 to 1	0.05 to 0.075
2	Slight	Cracks filled easily. Redecoration probably required. Several slight fractures inside building. Exterior cracks visible, some repointing may be required for weather tightness. Doors and windows may stick slightly.	1 to 5	0.075 to 0.15
3	Moderate	Cracks may require cutting out and patching. Recurrent cracks can be masked by suitable linings. Tack-pointing and possibly replacement of a small amount of exterior brickwork may be required. Doors and windows sticking. Utility services may be interrupted. Water tightness often impaired.	5 to 15 or a number of cracks greater than 3	0.15 to 0.3
4	Severe	Extensive repair involving removal and replacement of sections of walls, especially over doors and windows. Windows and door frames distorted. Floor slopes noticeable. Walls lean or bulge noticeably; some loss of bearing in beams. Utility services disrupted.	15 to 25 but also depends on number of cracks	Greater than 0.3
5	Very severe	Major repair required involving partial or complete reconstruction. Beams lose bearing; walls lean badly and require shoring. Windows broken by distortion. Danger of instability.	Usually greater than 25 but depends on number of cracks	—

8.7 HALF-TUNNELS FOR ROADS

Half-tunnels are excavated as overhangs of hard rock slopes along hill roads with one wall on the hill side and no wall on the valley side (Fig. 8.7). These half-tunnels have existed since 1980, despite no support of the roof in the middle and higher Indian Himalayas. Half-tunnels, which are excavated as overhangs of hill slopes, are superior to conventional full tunnels or open road excavations because they involve less cost and time. However, due to a lack of focus and their uncommon occurrence, the domain of half-tunnels has remained by and large unexplored. A photograph of a half-tunnel in Figure 8.7 shows how it looks exactly.

In addition to economy of construction, half-tunnels are attractions to tourists and help in the preservation of an ecosystem due to minimum disturbance to the slopes. Anbalagan et al. [16] have reported a detailed study. Figure 8.8 indicates the stability of a wedge above a half-tunnel. Span B_{ht} is found to be nearly equal to $1.7Q^{0.4}$ meters from limited field observations.

8.8 ROAD TUNNELS

8.8.1 General

The latter half of the 20th century has been called the dawn of the golden era of tunneling all over the world. Nothing succeeds like success. Since the early 1960s, many deep and long road and rail tunnels have been excavated through the Alps and the Rocky mountains. The 34-km-long Loetschberg tunnel is now the longest overland tunnel. The 58-km-long Gotthard tunnel, running parallel

FIGURE 8.7 Photograph of a half-tunnel [16].

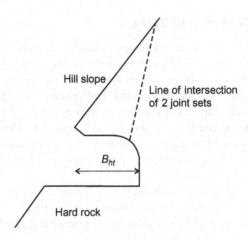

FIGURE 8.8   Half-tunnel along hill roads in hard rocks.

to the Loetschberg tunnel, will be the world's longest tunnel when it is completed by 2020. Undercity bypass tunnels along highways are becoming popular as in Australia. Several bypass road tunnels are being made to avoid complex landslides and dangerous debris flows. Jammu-Srinagar road with 11 long tunnels is under construction in Himalaya India. The experiences of tunneling in the tectonically disturbed, young and fragile Himalayas are precious for underground space engineers all over the world.

Development of a hill highway system should take into account the need for:

- Efficient transport shortening road routes
- Modest construction and maintenance costs
- High safety standards
- Preservation of the environment, monuments, and hill villages
- Operating in all seasons in case of strategic hill roads

The cost of road tunnels per kilometer of length may be less expensive than bridges per kilometer in hard rocks. In many tunnels the cost of permanent support amounts to 40 to 50% of the total cost. Steel fiber-reinforced shotcrete (SFRS) is slightly costlier but saves time. The escalation in cost may be as high as 80 to 200% in weak rock masses with geological surprises. The smooth underground bypass tunnels are constructed below congested crossings of roads or rail lines or too-busy highways in mega cities. They are more popular among senior citizens and disabled persons than superstructure passages without escalators.

The large traffic volume can be reduced by the development of long road tunnels below mega cities, in addition to the rail metro system. A highway system is the lifeline of a nation. It is key to rapid economic development. Hence many corporations are making huge investments in building infrastructure in developing nations. The metro system should be connected to railway and bus stations and airports for efficient, tension-free transport.

8.8.2 Traffic Safety

Research in Norway and international experience led to the following conclusions [17,18].

- The accident frequency of tunnels was 0.29 compared with 0.43 accidents per million vehicle-kilometers for open roads between tunnels; tunnels are safer than adjoining roads.
- Urban tunnels have a higher accident rate than tunnels in rural areas, but not higher than that on open roads.
- The tunnel entrance zone is far more dangerous (three times more risky in cold regions) than the interior of the tunnel (due to ice on the roadways).
- Good illumination increases safety.
- Wide variations exist in accident rates.
- Accidents occur more often in curved tunnels.

A wide road with gentle curves is considered to be much safer than a narrow hill road with sharp bends. If the speed level is reduced to about 50 kph, the geometry of the road is of minor importance in hills. Clean reflectors along the center line seem to encourage drivers to drive close to the tunnel walls for more safety.

8.8.3 Construction Details

The tunnel should be excavated more than the planned size to accommodate tunnel closures, which can be quite high in the case of squeezing and swelling grounds. Figure 8.9 shows a variation of typical costs (in 1999 US $) of tunnels per meter with the tunnel width. These costs are for excavation and support only and do not include the cost of final concrete lining or tunnel fittings (ventilation, lighting, rail lines, road pavement, etc.). These costs will naturally vary from nation to nation and with cost escalation due to geological surprises and poor management conditions. It is interesting to see that the cost of tunnel per meter increases linearly with the span of tunnels in a geological condition. Barton [19] has suggested underground construction costs with Q values for different diameters of tunnels (Figs. 8.10 and 8.11).

The cost of preinjection in tunnels to make it dry and to improve the quality of the rock mass has been studied by Barton [19], as suggested in Table 8.7.

Construction engineers express their desire of no rock bolting for rapid rate of tunneling on highways and for life of the support system of 100 years. Thus steel fiber-reinforced shotcrete is the right choice without rock bolts and steel ribs. The loose pieces of rocks should be scraped thoroughly before shotcreting for stronger bonding between two surfaces. The life span

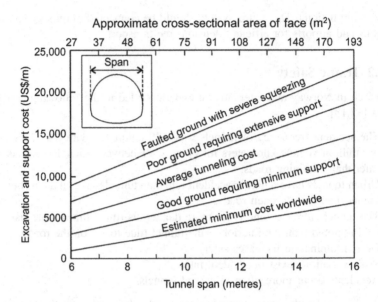

FIGURE 8.9 Approximate costs for tunnel excavation and support (in US$) (these costs do not include concrete lining, tunnel fittings, or tunnels driven by TBMs) [20].

of SFRS may be taken the same as that of concrete in polluted environment, which is about 50 years. The life may be increased to 60 years by providing an extra 50-mm cover of SFRS. If SFRS is damaged later, the corroded part may be scraped and a new layer of shotcrete may be sprayed to increase its life. The Norwegian Road Authority recommends a minimum thickness of 80 mm for road tunnels. Further, the width of the pillar between twin road tunnels is more than the sum of half-widths of adjoining openings in non-squeezing grounds. The width of pillars is also more than the total height of the larger of two tunnels.

Concrete instead of asphalt pavements are generally preferred inside tunnels. Minimum thickness is 15 cm and increases with higher traffic volumes.

Natural ventilation occurs as a result of a temperature/pressure difference between two ends of short tunnels, which is caused by a difference in the climatic conditions. Longitudinal ventilation with fans mounted on the crown is by far the most commonly used in short highway tunnels. The distance between fans should not be less than 80 meters. The main objectives of any ventilation system are to ensure a safe level of poisonous gases (CO, NO_2, dust) and to reduce unpleasant smells. In tunnels longer than 2 km, vertical ventilation shafts are provided in the shallow rock cover, such as in Hoyanger tunnel, Norway. Henning [22] has presented the design details of the ventilation system. The risk of blast or fire inside a tunnel should also be considered in the ventilation design.

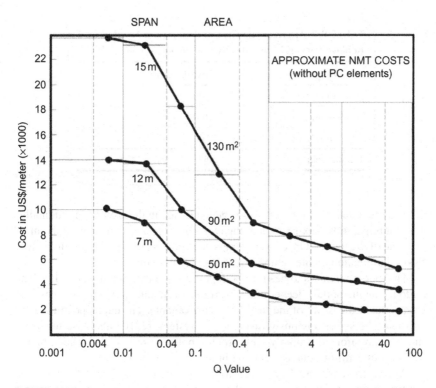

FIGURE 8.10 Cost of tunneling in various rock classes represented by Q value for different diameter tunnels [19].

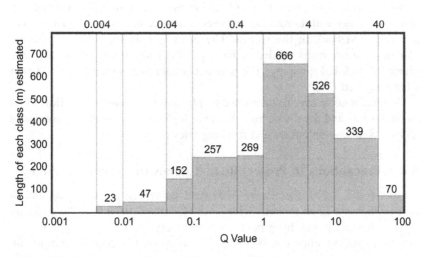

FIGURE 8.11 Length of tunneling in various rock classes represented by Q value for obtaining Figure 8.10 [19].

TABLE 8.7 Approximate Cost of Preinjection Needed to Achieve Various Levels of "Dryness" in 90-m^2 Tunnels [21]

Inflow (approx.)	Cost (US$/m)
20 liter/min/100 m	1400
10 liter/min/100 m	2300
5 liter/min/100 m	3500
1–2 liter/min/100 m	≈5000

The illumination of rock tunnels is generally along the crown longitudinally. The illumination is maximum near the two entrances and minimum in the middle zone of the tunnel. Once a year lamps and fittings are cleaned and inspected. Good illumination creates emotions of happiness among drivers.

The modern trend is to beautify both portals to entertain onlookers. It is better to locate the portals deeper into the ground or mountain where rock cover at least equal to the width of the road tunnel is available. The cut slope of the portal should be stable obviously (using Q_{slope} system; [23]). Otherwise the same should be reinforced properly with grouted bolts. Alternatively, a thick breast wall (about 1 m) of concrete may be built to ensure its stability.

Approach roads/rail lines need to be widened sufficiently for tackling boulder jumping or snow avalanches. Catch drains of a proper depth and width should be made on both sides of the road, according to the height and slope of cuts and sizes of boulders on the slope, if any. A fence of 3.5 m height should, therefore, be erected along both drains and tied to poles at about 2 m center to center with horizontal bracings at 1 m center to center. Then poles are anchored back in the slopes. The price is paid back if the wire net (4-mm-diameter wires welded at 10 × 10 cm or alternative) withstands the impact of rock fall /jumping. The wire net should than be replaced soon after where required [24].

Vehicular noise may be dampened by planting trees, shrubs, and flowers on embankments and the divide zone between highways, generating emotions of happiness among onlookers and reducing rates of accidents.

8.8.4 Precautions to Protect Road Tunnels from Deterioration

As of April 1, 1990, there were 6705 road tunnels in use in Japan, almost all of which are lined tunnels, with a total length of 1970 km. The majority of road tunnels in use have been built relatively recently: nearly half have been constructed since the 1970s. In recent years, the deterioration of linings resulting from cracks and deformation has been found in some road tunnels [25].

Forms of deterioration include cracks (some so small that they could only be seen from inside the sideway), deformation, piece separation and massive falling off from the tunnel, road heaving, and deformation and collapse of side drains. In a broader sense, deterioration also includes water leakage. Because so many road tunnels are affected by water leakage, tunnels where the lining is still sound, despite water leakage, are not considered to be deteriorated [25].

The New Austrian Tunneling Method (NATM), which is currently the standard method of constructing mountain tunnels, almost completely prevents water inflow by the use of waterproof measures such as waterproof sheets installed between the lining and the shotcrete. However, the conventional steel support and timbering method and other methods used before the introduction of NATM make treatment of water inflow behind the lining very difficult, which has increased the number of tunnels with water inflow.

Following are highlights of a study on deterioration of road tunnels in Japan [25].

- About 60% of road tunnels currently in service suffer from water leakage, whereas 24% suffer from deterioration other than leakage. Cracks in linings are the most common form of deterioration.
- There is no clear correlation between the occurrence of deterioration and the number of years in service. Deteriorated tunnels were constructed predominantly between 1950 and 1980. The ratio of deteriorated tunnels to the total number of tunnels constructed in each service year group is larger for tunnels constructed between 1930 and 1980 than for those put into service before 1930 and after 1980.
- Deterioration occurred within 30 years of use in 90% of tunnels and after 30 years of use in 10% of tunnels.
- Field personnel cited "water leakage and frost damage" and "superannuation" as the two most common causes of deterioration.
- "Extrusion," "road surface deformation," and "side drain deformation" tended to occur relatively quickly after tunnels were built, whereas "separation," "massive falling," and "separating out of lime, etc." tended to occur after many years.

8.8.5 Cost of Construction

In 1984, a survey was launched into the cost of 33 two-way road tunnels. The longest of these was Saint-Gotthard, which is a 16,918-m-long road tunnel. For road tunnels more than 2 km long, a close correlation was found between cost (C) and length (L), using the following formula:

$$C = a L^{1.36} \tag{8.1}$$

Using this formula, a 15-km-long tunnel would cost 1.8 times, not 1.5 times, more than a 10-km-long tunnel.

The rapid cost increments are partly due to ventilation needs because power needs increase more quickly with the length to be ventilated and require shafts and intermediate ventilation points. For a rail tunnel, costs would probably increase less quickly and would vary more or less linearly with the length of the tunnel [26].

8.9 SUBSEA TUNNELS

8.9.1 General

Civil engineers have performed miracles by tunneling below deep oceans to link the hearts of the people of different nations. Heijboer and colleagues [3] have presented a detailed case history of the (twin, 12 m apart) Westerschelde tunnels of 6.6 km length and 11 m diameter in clayey soil and 12–60 m below the ocean (tidal river) in The Netherlands. A slurry shield TBM was chosen due to the presence of both noncohesive and cohesive soils at this site and bentonite slurry was used. The cost of the project will be recovered by toll collection in about 30 years. The circular tunnel lining consists of 2-m-wide R.C.C. rings of thickness 0.45 m and compressive strength of 55 MPa with seven segments and a keystone, surrounded by a neoprene lining for waterproofing. A fire-resistant cladding of 27 mm thickness has been sprayed on the inner surface of the aforementioned lining to withstand fire at 1350°C for 2 hr. The life of the lining of concrete is expected to be more than 100 years. There are 26 cross-connections between the twin tunnels for escape routes to reduce chances of individual accidents to 1/1,000,000; maximum speed is 80 kph, and two separated tunnels give no chance for head-on collisions of vehicles.

Martin [27] has reported a case history of an undersea Vardo tunnel in Norway. The undersea tunnel is 2800 m long, 8 m wide, and 87 m below sea. The tunnel passes through shales and sandstones with a thick fault zone. The whole fault zone is lined with concrete, but the rock tunnel is supported with shotcrete and bolts without lining. The cost of the undersea tunnel was estimated as slightly less (about 10%) than the cost of the overhead bridge. Because water leakages have been small, the rock mass was grouted.

Table 8.8 lists 25 Norwegian subsea tunnels to replace the ferries on the sea. In Norway, eight subsea tunnels have been built for oil industries and another eight for water supply and sewage (Table 8.9). All projects are financed with fees. Figure 8.12 shows a typical cross section of a subsea tunnel. The minimum rock cover should exceed 50 m in subsea tunnels [28]. The thickness of SFRS is above 70 mm in subsea tunnels. The rock bolts have extensive protection for corrosion. The rock mass is grouted all around the tunnels to withstand water pressure due to the sea (with grouting pressure of about 10 MPa). Thus the life span of subsea tunnels is expected to be from 15 to 40 years.

Culverwell [29] has compiled Table 8.10 of 67 submerged (immersed) tube-road and rail line tunnels. It is thus seen that submerged tunnels are used extensively on road and rail lines. Table 8.10 may help in planning short submerged tunnels.

TABLE 8.8 Norwegian Subsea Road Tunnels Completed or Under Construction [28]

Serial No.	Project	Year Completed	Main Rock Type	Cross Section, m²	Total Length, km	Minimum Rock Cover, m	Lowest Level, m Below Sea
1	Vardo	1981	Shale, sandstone	53	2.6	28	88
2	Ellingsoy	1987	Gneiss	68	3.5	42	140
3	Valderoy	1987	Gneiss	68	4.2	34	145
4	Kvalsund	1988	Gneiss	43	1.6	23	56
5	Godoy	1989	Gneiss	52	3.8	33	153
6	Hvaler	1989	Gneiss	45	3.8	35	121
7	Flekkeroy	1989	Gneiss	46	2.3	29	101
8	Nappstraumen	1990	Gneiss	55	1.8	27	60
9	Fannefjord	1991	Gneiss	54	2.7	28	100
10	Maursund	1991	Gneiss	43	2.3	20	92
11	Byfjord	1992	Phyllite	70	5.8	34	223
12	Masrafjord	1992	Gneiss	70	4.4	40	132
13	Freifjord	1992	Gneiss	70	5.2	30	132
14	Hitra	1994	Gneiss	70	5.6	38	264
15	Tromsoysund	1994	Gneiss	60[a]	3.4	45	101

(Continued)

TABLE 8.8 Norwegian Subsea Road Tunnels Completed or Under Construction [28]—cont'd

Serial No.	Project	Year Completed	Main Rock Type	Cross Section, m²	Total Length, km	Minimum Rock Cover, m	Lowest Level, m Below Sea
16	Bjoroy	1996	Gneiss	53	2.0	35	85
17	Sloverfjord	1997	Gneiss	55	3.3	40	100
18	Norh Cape	1999	Shale, sandstone	50	6.8	49	212
19	Oslfjord	2000	Gneiss	79	7.2	32[b]	134
20	Froya	2000	Gneiss	52	5.2	41	164
21	Ibestad	2000	Micaschist, granite	46	3.4	30	125
22	Bomlafjord	2000	Greenstone, gneiss, phyllite	74	7.9	35	260
23	Skatestraumen	2002	Gneiss	52	1.9	40	80
24	Halsnoy	2008	Gneiss	50	4.1	45	135
25	Eiksundet	2007	Gneiss, gabbro, limestone	71	7.8	50	287

[a]The only tunnel with two tubes.
[b]Assumed rock cover from site investigations, proved to be lacking at deepest point.

TABLE 8.9 Some Main Norwegian Subsea Tunnels for Water, Gas, and Oil [28]

Serial No.	Project	Year Completed	Main Rock Type	Cross Section, m²	Total Length, km	Minimum Rock Cover, m	Lowest Level, m Below sea
1	Frierfjorden, gas pipeline	1976	Gneiss and claystone	16	3.6	48	253
2	Karsto, cooling water	1983	Phyllite	20	0.4	15	58
3	Karmsund (Statpipe), gas pipeline	1984	Gneiss and phyllite	27	4.7	56	180
4	Fordesfjord, (Statpipe)	1984	Gneiss	27	3.4	46	160
5	Forlandsfjord, (Statpipe)	1984	Gneiss and phyllite	27	3.9	55	170
6	Hjartoy, oil pipeline	1986	Gneiss	26	2.3	38 (6 m at piercing)	110
7	Kollsnes (Troll), gas pipeline	1994	Gneiss	45–70	3.8	7 m at piercing	180
8	Karsto, new cooling water	1999	Phyllite	20	3.0/0.6	a	60,10
9	Snohvit, water intake/outlet	2005	Gneiss	22	1.1/1.3	a	111/54
10	Aukra, water intake/outlet	2005	Gneiss	20/25	1.4/1.0	5/8 (5.5 m at piercing)	86/87

a No information.

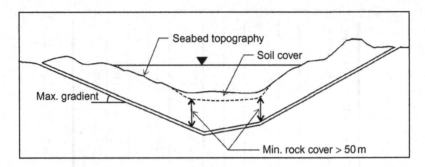

FIGURE 8.12   Typical section of subsea tunnel [28].

A tunnel is being planned between London and New York City in the Atlantic Ocean. Another 70-km-long tunnel is contemplated to join Russia with the top of Canada. Thus, the future is fascinating and full of potential for a world civilization.

8.9.2 Maintenance

The soundness of the tunnel structure must be assessed when performing maintenance [30–32]. Specifically, soundness is assessed based on data obtained for the condition of the ground and lining concrete. Figure 8.13 indicates the maintenance plan and procedure for the 54-km-long Seikan undersea tunnel in Japan and items to be surveyed [28].

Based on observations made during the maintenance supervision of the Seikan tunnel in Japan, the following results were highlighted [33].

- The undersea tunnel structure remains in a good condition after 14 years of operation.
- Although the volume of inflow water increases right after a large earthquake, it is gradually decreasing. The fluctuation of inflow water seems to be attributed to changes of the amount of inflow water passing through fissures in rocks.
- No change in the inclusions of inflow water was found.
- The impact (acceleration) of an earthquake on the undersea tunnel is much less than that on structures on the ground surface.
- Although products of deterioration generated through reactions between seawater and cement ingredients in the grout materials were not found, monitoring should be continued to determine if there is a change in composition of the grout materials.
- The change in tunnel cross sections was minimal, and no abnormalities were detected in lining concrete.
- Although changes in the strain on lining concrete of the main tunnel occur in a cyclic manner, the strain has a minimal impact on the integrity of lining concrete.

TABLE 8.10 Immersed Tube Road and Rail Tunnels [29]

Serial No.	Year	Name	Purpose	Location	Tube Length, m	Lanes Tracks	Form
1	1910	Detroit River	Railway	Michigan, Ontario Canada	800	2 × 2	S
2	1914	Harlem River	Railway	New York	329	4 × 1	S
3	1927	Freidrichshafen	Pedestrian footway	Berlin	120	—	R
4	1928	Posey	Road	Oakland California	742	2	R
5	1930	Detroit Windsor	Road	Michigan, Ontario Canada	670	2	S
6	1940	Bankhead	Road	Mobile Alabama	610	2	S
7	1941	Mass	Road	Rotterdam Netherlands	587	2 × 2	R
8	1942	State Street	Railway	Chicago Illinois	61	2 × 1	S
9	1944	Aji River	Road	Osaka Japan	49	2 × 1	S
10	1950	Washburn	Road	Pasadena Texas	457	2	S
11	1952	Elizabeth River (1)	Road	Portsmouth Virginia	638	2	S
12	1953	Baytown	Road	Baytown Texas	780	2	S
13	1957	Baltimore	Road	Baltimore Maryland	1920	2 × 2	S
14	1957	Hampton Roads (1)	Road	Virginia	2091	2	S
15	1958	Havana	Road	Cuba	520	2 × 2	R
16	1959	Deas Island	Road	Vancouver Canada	629	2 × 2	R

(Continued)

TABLE 8.10 Immersed Tube Road and Rail Tunnels [29]—cont'd

Serial No.	Year	Name	Purpose	Location	Tube Length, m	Lanes Tracks	Form
17	1961	Rendsburg	Road	Keil West Germany	140	2 × 2	R
18	1962	Webster Street	Road	Oakland California	732	2	R
19	1962	Elizabeth River (2)	Road	Portsmouth Virginia	1056	2	S
20	1964	Chesapeake Bay (a) Thimble Shoal Tunnel (b) Baltimore Channel Tunnel	Road	Virginia	(a)1750 (b)1661	2	S
21	1964	Liljeholmsviken	Railway	Stockholm	123	2	R
22	1964	Haneda	Road	Tokyo	56	2 × 2	S
23	1964	Haneda	Mono rail	Tokyo	56	2	S
24	1966	Coen	Road	Amsterdam	540	2 × 2	R
25	1967	Benelux	Road	Rotterdam Netherlands	745	2 × 2	R
26	1967	Lafontaine	Road	Montreal Canada	768	2 × 3	R
27	1967	Vieux-Port	Road	Marseilles France	273	2 × 2	R
28	1967	Tingstad	Road	Gothenburg Sweden	452	2 × 3	R
29	1968	Rotterdam Metro	Railway	Rotterdam Netherlands	1040	2 × 1	R
30	1969	IJ	Road	Amsterdam	790	2 × 2	R
31	1969	Scheldt E3 (JFK Tunnel)	Road/railway	Antwerp Belgium	510	2 × 3 road 2TR	R
32	1969	Heinenoord	Road	Barendrecht Netherlands	614	2 × 3	R

33	1969	LIMF Jord	Road	Arlborg Denmark	510	2 × 3	R
34	1969	Parana (Hernandias)	Road	Santafe Argentina	2356	2	R
35	1969	Dojima River	Railway	Osaka Japan	72	2 × 1	R
36	1969	Dohtonbori River	Railway	Osaka Japan	25	2 × 1	S
37	1969	Tama River	Railway	Tokyo	480	2 × 1	S
38	1970	Keihin Channel	Railway	Tokyo	328	2 × 1	S
39	1970	Bay Area Rapid Transit	Railway	San Francisco California	5825	2 × 1	S
40	1971	Charles River	Railway	Boston Massachusetts	146	2 × 1	S
41	1972	Cross-Harbour Tunnel	Road	Hong Kong	1602	2 × 2	S
42	1973	East 63rd ST Tunnel	Railway (part)	New York	2 × 229	4 × 1	S
43	1973	110	Road	Mobile Alabama	747	2 × 2	S
44	1973	Kinuura Harbour	Road	Handa Japan	480	2	S
45	1974	Ohgishima	Road	Kawasaki Japan	664	2 × 2	S
46	1975	Elbe	Road	Hamburg Germany	1057	3 × 2	R
47	1975	Vlake	Road	Zeeland Netherlands	250	2 × 3	R
48	1975	Sumida River	Railway	Tokyo	201	2 × 1	S
49	1976	Hampton Roads (2)	Road	Virginia	2229	2	S
50	1976	Paris Metro	Railway	Paris	128	2	R
51	1976	Tokyo Port	Road	Tokyo	1035	2 × 3	R

(Continued)

TABLE 8.10 Immersed Tube Road and Rail Tunnels [29]—cont'd

Serial No.	Year	Name	Purpose	Location	Tube Length, m	Lanes Tracks	Form
52	1977	Drecht	Road	Dordrecht Netherlands	347	4 × 2	R
53	1978	Prinses Margriett	Road	Sneek Netherlands	77	2 × 3	R
54	1978	Kil	Road	Dordrecht Netherlands	330	2 × 3	R
55	1979	Washington Channel	Railway	Washington DC	311	2 × 1	S
56	1979	Kawasaki	Road	Kawasaki Japan	840	2 × 3	S
57	1979	Hong Kong Mass Transit Railway	Railway	HongKong	1400	2 × 1	R
58	1980	Hemspoor	Railway	Amsterdam	1475	3 × 1	R
59	1980	Botlek	Road	Rotterdam Netherlands	508	2 × 3	R
60	1980	Daiba	Railway	Tokyo	670	2 × 1	S
61	1980	Dainikoro	Road	Tokyo	744	2 × 2	R
62	1982	Rupel	Road	Boom Belgium	336	2 × 3	R
63	1984	Spijkenisse	Railway	Rotterdam	530	2 × 1	R
64	1984	Coolhaven	Railway	Rotterdam	412	2 × 1	R
65	1984	Keohsiung Harbour	Road	Taiwan	1042	2 × 2	R
66	1985	Fort McHenry	Road	Baltimore	1638	4 × 2	S
67	1986	Elizabeth River (3)	Road	Virginia	762	1 × 2	S

Note: 1. Year is that of completion. 2. Denotes part of underground railway system. 3. Form of tunnel denotes thus: S-steel shell; R-reinforced or prestressed concrete box.

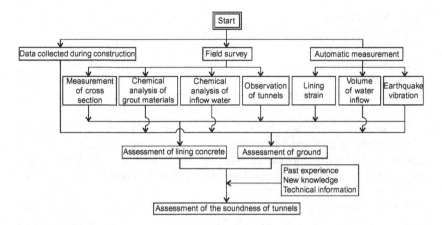

FIGURE 8.13　Flow of tunnel maintenance planning [33].

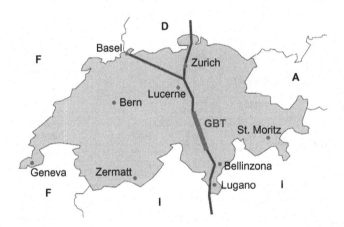

FIGURE 8.14　Location of the Gotthard Base Tunnel in Alp Transit Railway Project.

8.10 CASE HISTORIES

8.10.1 Gotthard Base Tunnel (GBT)

The 57-km-long GBT is the longest among the base tunnels to be built as part of Switzerland's Alp Transit Railway Project. It leads from the north via Zurich into the south of Switzerland and to Milan in Italy (Fig. 8.14).

Planning and Design

Preliminary work for the GBT began in 1996, with excavation of access tunnels and shafts. The main construction work began in 2002 and will be completed in

2015. After completion of the railway installations and commissioning, the GBT will start commercial operation at the end of 2017.

Routing of the GBT had to be accomplished in such a way as to allow high-speed passenger trains to travel at 250 km/h and freight trains at 140 km/hr. As a consequence, maximum gradients of 1% are permitted, and the portals must be located roughly 600 to 800 m lower than the portals of the existing Gotthard railway tunnel.

The geological conditions had to be taken into account, especially the attempt to avoid difficult geological formations or to pass through zones with minimal expansion. Further considerations were the overburden as well as ensuring a safe distance to existing reservoirs. The outcome was a slightly z-shaped horizontal routing of the tunnel (Fig. 8.15).

With the northern portal near Erstfeld and the southern one near Bodio, the tunnel is just under 57 km in length. The complete tunnel system consists of 153.3 km of access tunnels, shafts, railway tunnels, connecting galleries, and auxiliary structures (Fig. 8.16).

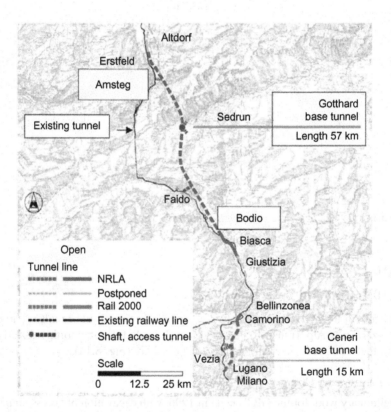

FIGURE 8.15 Plan view of the Gotthard Base Tunnel.

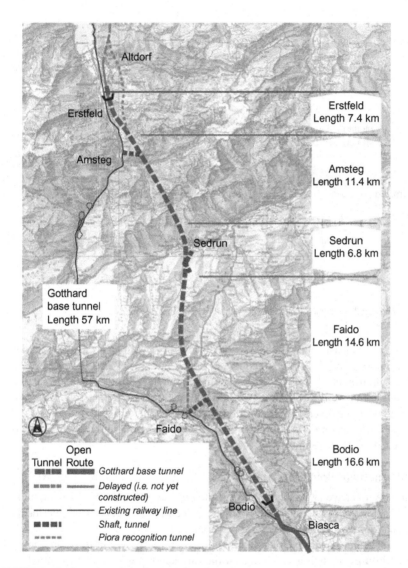

FIGURE 8.16 Overview of the Gotthard Base Tunnel.

Tunnel System

Following intensive investigations into possible tunnel system and planning parameters (such as construction operations, capacity, maintenance of facilities, safety, and, last but certainly not least, construction and operational costs, as well as risks pertaining to construction costs and duration), it was decided to select a system comprising two single-track tunnels running parallel to one another.

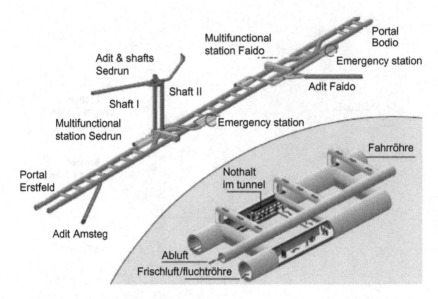

FIGURE 8.17　Scheme of the tunnel system.

The GBT consists of two parallel single-track tubes with an excavation diameter varying from 8.8 to 9.5 m linked by cross-passages approximately every 312 m. Two multifunction stations (MFS) are located in the Sedrun and Faido sections, one-third and two-thirds along the length of the tunnel, respectively. These will be used for the diversion of trains to the other tube via crossovers, to house technical infrastructure and equipment, and as emergency stopping stations for the evacuation of passengers (Fig. 8.17).

Intermediate Points of Attack

In order to arrive at an acceptable construction period for the 57-km-long tunnel, as well as to provide adequate ventilation when finally operational, intermediate points of attack (adits) have to be created.

These intermediate points (adits) have been set up within the topographical possibilities in such a way that, on the one hand, they divide the tunnel into roughly equal sections and, on the other hand, permit zones, which are tricky and need to be tackled at the earliest possible stage.

The construction program and cost estimations with a large number of concepts relating to the points of attack led to the following intermediate points of attack (adits) being selected for the GBT in the preliminary project.

- *Erstfeld*: northern portal.
- *Amsteg*: a horizontal, approximately 1.2-km-long access tunnel.
- *Sedrun*: via a blind shaft with a depth of 800 m and a clear diameter of 8.0 m opened up through an approximately 1-km-long horizontal access tunnel.

An additional inclined shaft connection from the head of the shaft to the surface serves as ventilation when operational.

- *Faido*: an inclined tunnel with a gradient of approximately 12%, which is roughly 2.7 km in length and overcomes a 330-m difference in height.
- *Bodio*: southern portal.

The Sedrun and Faido sections each include a MFS.

Design Criteria for the Shaft

The 800-m-deep vertical shaft is the only access to the four tunnel headings starting from the bottom of the shaft. Therefore, the layout of the shaft has to consider the following points:

- Transportation of materials in the tunnel: such as machines, concrete, and other lining support material.
- Transport of muck material from the headings through the shaft and through the access tunnel to the surface.
- Fresh air for tunnel ventilation. (fresh air in/waste air out).
- Additional pipes for water supply, compressed air, electrical supplies, communications, and so on.
- The safety of the workers in all possible conditions has to be guaranteed.

Standard Profile Design

The standard profile of the tunnel tubes is currently based on a two-shell support system, which is secured with shotcrete and reinforced with wire mesh, rock bolts, and, if necessary, steel arches and a membrane seal. In the TBM-driven sections, excavated diameters vary between 8.8 and 9.5 m (Fig. 8.18).

Construction Method

Tunnel geology is marked by three major gneiss complexes over 53 km (Aar-Massif, Gotthard-Massif, and Penninic Gneiss zone), which can largely be regarded as favorable formations for construction purposes. In zones with a high overburden, foliations are found in an almost perpendicular position and will be crosscut at a right angle during tunneling. The Tavetsch Intermediate Massif and several younger sedimentary zones, which are located in between, represent the technically difficult section of the tunnel. The overburden will be very high over major sections of the tunnel. In fact, it amounts to more than 1000 m for roughly 35 km to more than 1500 m over 20 km and even over 2000 m over approximately 5 km. The maximum overburden is about 2300 m (Fig. 8.19).

In the entire GBT, the two zones of the Piora Basin and the Tavetsch Intermediate Massif are regarded as the most difficult zones to drive through. One of the most important measures for mastering the hazards in tunneling is selection of the appropriate excavation method.

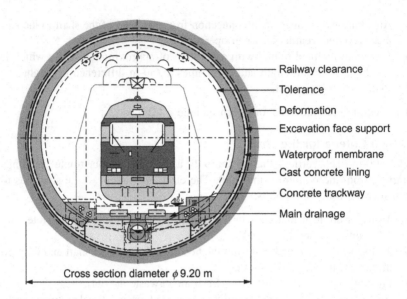

FIGURE 8.18 The TBM-driven cross section of the Gotthard Base Tunnel.

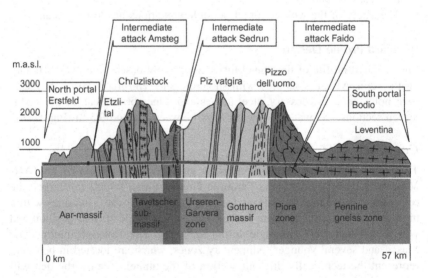

FIGURE 8.19 Geological longitudinal profile.

There is no general rule for selection of the excavation method. The excavation method has to be selected based on project-specific criteria depending on the requirements of the project. Project requirements may be:

- Health and safety
- Environmental aspects

- Aspects of future operation
- Design (including schedule and costs)
- Construction (including schedule and costs)
- Third parties and existing facilities, including buildings, bridges, tunnels, roads, surface and subsurface railways, pavements, waterways, flood protection works, surface and subsurface utilities, and all other structures/infrastructure that can be affected by execution of the works

In the case of the GBT, the excavation method has to be determined for the five main construction sections with lengths from 6.5 km up to nearly 15 km.

In the central construction section of Sedrun, tunnel excavation has to be executed in the northward and southward directions starting from the bottom of an 800-m-deep shaft. Geological conditions were predicted with a high variance from very good hard rock to very poor rock with a high squeezing potential and an overburden of 1.0 km; additionally, karstic rock zones were expected, and the tunnel in the southward drive has to be excavated close to a concrete arch dam. The northern drive was foreseen with a length of 2.15 km and the southern drive with a length of 4.6 km.

For the other four construction sections of Erstfeld (7.1 km), Amsteg (11.4 km), Faido (12.2 km), and Bodio (14.8 km), the boundary conditions were not so restrictive. Therefore, mainly the restrictions of construction time and costs determined the excavation method. In the GBT, around 65% of the entire tunnel system is being excavated by tunnel boring machines. Around 35% of the total length, mainly access tunnels, main tunnels in the central construction section of Sedrun, and the multifunction station at Faido, is being driven by the conventional drill-and-blast tunneling method (Fig. 8.20).

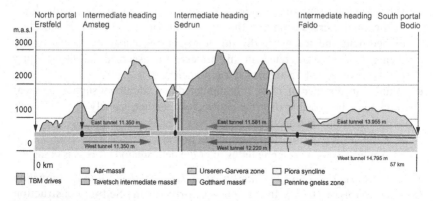

FIGURE 8.20 Scheme of the excavation method along the Gotthard Base Tunnel.

Conventional Tunneling Method

Excavation of a 1-km-long zone of squeezing rock (Sedrun Northern Drive). The northern drive of the Sedrun section has to cross the Tavetsch Intermediate Massif North. This intermediate massif underwent very strong tectonic deformation during formation of the Alps. According to more than 3-km-long exploratory drilling, squeezing rock conditions were expected to be encountered in this zone. Therefore, the excavation of zones with squeezing rock in the Sedrun construction section is based on the following principles.

- The profile will be excavated on (the static force) an ideal circular shape with an additional space for deformations of up to 70 cm.
- The rock support elements should be able to sustain initial large deformations.
- For the first rock support, only highly deformable materials with a plastic failure mode will be allowed, such as steel ribs with special sliding joints and rock anchors. Ground support will be comparatively low in the first phase (yielding phase).
- Support materials with a brittle failure mode such as shotcrete will be placed only when deformations come practically to an end. In this phase (resistance phase), support forces will be elevated to the necessary level for a stable equilibrium.

The aim was to find a construction procedure with an immediate ring closure. Only full-face excavation allows the fastest possible ring closure time. The procedure for closing the ring was as follows.

- Full face excavation in a circular shape (80 to 135 m^2) in 1.3-m steps.
- Placement of steel ribs after each advance (2 x TH 44, at center-to-center distances of 33, 50, 66, and 100 cm).
- Placement of 12-m-long self-drilling radial anchors after each round.
- Placement of 18-m-long self-drilling tunnel-face anchors after every 6 m of advance.
- When the full length of the sliding joint has been taken up, the circular force in the steel increases and so the risk of buckling increases. For perfect embedment and to ensure the full load-bearing capacity, steel ribs are then completely shotcreted. In this project, it is assumed that the yielding phase is completed 75 m behind the tunnel face.
- Within this distance of 75 m, it is stipulated that additional supporting measures must be possible over the entire cross section, namely
  - Insertion of additional steel rings
  - Reanchoring during the deformation process
  - Application of a shotcrete lining after the deformation process is complete

Tunnel construction close to concrete arch dam with conventional tunneling (Sedrun Southern Drive). In the southern drive of the Sedrun construction section, the GBT passes through a distance of a little more than 1 km under the 127-m-high concrete arch dam at Nalps. The main hazard scenario is the occurrence of unacceptable surface deformation due to dewatering effects in the jointed rock mass during and after tunnel construction. It is quite clear that grouting work can be carried out much more easily in conventional tunneling than from a TBM, where the working space is very limited.

Mastering an extended fault zone during construction of the multifunction station at Faido. Multifunction station construction consists of underground emergency stations, service caverns, crossovers, connecting galleries, ventilation galleries, and shafts. Over a comparatively short distance, all kinds of different cross sections from 30 m^2 up to over 200 m^2 have to be excavated. This type of structure is a typical case for application of the conventional tunneling method.

Tunnel Boring Machine Excavation Method

Bodio: lessons learned from a 12-km TBM drive. After a short time of starting the TBM excavation from Bodio, a horizontal fault zone was encountered inside the tunnel cross section and existed within the tunnel section for more than 500 m. The fault zone produced big overbreaks at the top and reduced the advance rate dramatically to around 3 m per working day. Another fault zone produced high deformations already in the area of the shield and also caused damage to the TBM in one case. Behind the shield, the deformation, and, in some places, the floor heaving became excessive. As there was no possibility to increase the bore diameter, the tunnel had to be reprofiled along a length of several hundred meters at a very high cost (Fig. 8.21).

Amsteg: lessons learned from an 11-km TBM drive. Two TBMs of a similar construction type started the excavation at Bodio in the southern direction. A fault zone of 8- to 15-m thickness was found in front of the cutter head; the whole zone around the cutter head had to be consolidated by grouting measures. The fault zone finally was consolidated with cement grouting from a niche in the

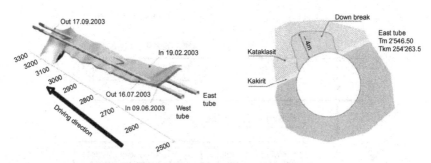

FIGURE 8.21 Horizontal fault zone at Bodio.

eastern tube. Grouting with gel protected the machine. Finally, more than 2800 m of drill holes for grouting was drilled, and 110 tonnes of cement and 50 tonnes of gel were used for grouting. Over the whole TBM drive in Amsteg, average advance rates were 11.5 m per working day (Fig. 8.22).

Faido. At the Faido section, difficulties began a short time after the start of excavation. The deformation behind the cutter head became locally large, and because there was a risk that backup installation could not pass through the excavated hole, the cutter head diameter changed from 8.80 to 9.40 m in order to have more space for deformations and rock support. The main elements for rock support are steel ribs, anchors, meshes, and shotcrete. Flexible supports attract less support pressure. A delay in installation of supports but within stand-up time also reduces the support pressure (Fig. 8.23).

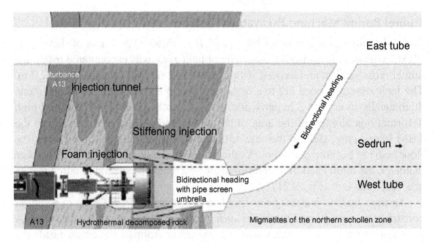

FIGURE 8.22   Overcoming a fault zone at Amsteg.

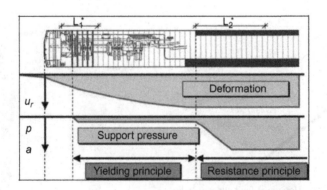

FIGURE 8.23   Faido: concept for the TBM drive at the end of the yielding phase.

Comparison between Two Different Excavation Methods

The excavation experience of the GBT shows significantly higher advance rates for long TBM drives than for conventional tunneling except in the case when isolated fault zones cause longer standstills for the TBM drive. A well-organized conventional drive allows mastering of large deformations and thereby mastering of high pressure in squeezing conditions.

The application of auxiliary construction measures for ground improvement (grouting), ground reinforcement (fore poling, face anchoring), and dewatering is much easier for conventional tunneling than for TBM excavation. A TBM drive is not technically and economically feasible in very difficult rock conditions, as demonstrated in Sedrun North and the Faido multifunction station.

Special Requirements during Deep Tunnel Construction

Deep tunnels not only have their specialties with regard to geotechnics and rock mechanics, but also in many other aspects:

High temperatures
Water inflow
High water pressures

For AlpTransit, rock temperatures of more than 40°C are anticipated, together with relatively high humidity. Waste heat from machinery and other factors will cause an additional increase in the temperature inside the tunnel. High temperatures have an additional negative effect on corrosive chemical attacks of dissolved minerals in the groundwater because they speed up all the relevant chemical reactions. The bearing parts of the tunnel structure need to be sealed off adequately to avoid major maintenance work during the life cycle of the tunnel.

8.10.2 Lötschberg Base Tunnel (LBT)

The Lötschberg base line has three tasks: for transit traffic, as an extension of the Swiss Rail 2000 system to the Valais, and as a link for the A-6 and A-9 motorways by means of a car on train cross service. The LBT runs from Frutigen in the Kandertal valley (Berner Oberland) to Raron in Valais. It is 34.6 km long and is designed as a twin-tube, unidirectional single-track rail tunnel (Fig. 8.24).

Planning and Design

The total length of the Lötschberg base tunnel is 34.6 km. To the south it links up with the existing Simplon line. The design speed varies with the type of train: 230 km/hr for high-speed passenger trains and 140 km/hr for shuttle trains.

The Lötschberg base tunnel is designed as a tunnel system with two separate single-track tubes. The distance between the two tubes varies from 40 to 60 m, depending on the quality of the rock mass. Transverse tunnels connect the tubes every 300 m. The cross-sectional diameter for excavation with a tunnel boring

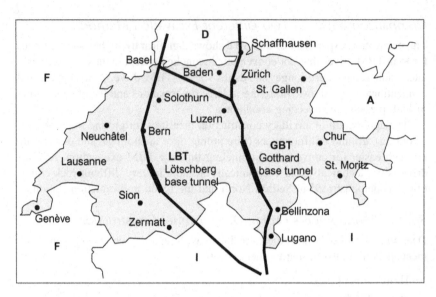

FIGURE 8.24   The Alp transit scheme.

machine is 9.43 m. Excavation with the traditional drill-and-blast method yields a surface area between 62 and 78 m^2.

Safety is a very important aspect in this project during planning and design; from results of the risk analysis, the new tunnel cannot be designed without adopting the following measures:

Measures to prevent accidents (appropriate conveys, railway well maintained)
Measures to reduce fatalities and damages
Measures to improve self-rescue opportunities
Measures to improve the possibilities of external rescue

According to the safety analysis, two emergency stations have already been built, one in Mitholz and the other in the vicinity of the village of Ferden, about 700 m underneath the ground level of the Lötschental valley. The emergency stations were designed for rescuing passengers from a train in case of fire by use of lateral adits, to remove smoke by a ventilation system, and to pump fresh air into the rescue galleries.

The northern portal at Tellenfeld near Frutigen is located at a height of 780 m, and the southern portal at Raron at 660 m. The tunnel rises by 0.3% until it reaches its apex in the mountains between the cantons of Berne and Valais and drops by roughly 1.1% at its southern end. The tunnel has five major construction sites at Frutigen and Raron portals and intermediate points of attack at Ferden (portal at Goppenstein Station), Steg (Niedergesteln portal), and Mitholz (Fig. 8.25).

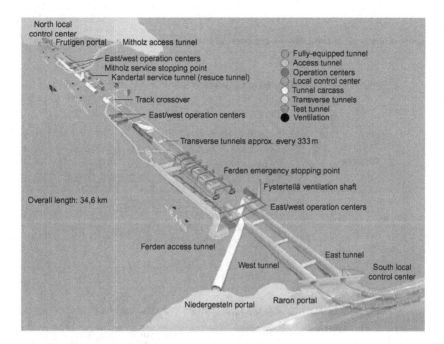

North local
control center
Frutigen portal ⋅ Mitholz access tunnel
East/west operation centers
Mitholz service stopping point
Kandertal service tunnel (resuce tunnel)

○ Fully-equipped tunnel
○ Access tunnel
◐ Operation centers
○ Local control center
○ Tunnel carcass
◐ Transverse tunnels
○ Test tunnel
● Ventilation

Track crossover

East/west operation centers

Transverse tunnels approx. every 333 m

Ferden emergency stopping point

Fystertellä ventilation shaft

Overall length: 34,6 km

East/west operation centers

Ferden access tunnel

East tunnel

West tunnel

South local
control center

Niedergesteln portal   Raron portal

FIGURE 8.25   Scheme of the LBT system.

Selection of the specific temporary and permanent support profile is decided at the tunnel face as a function of the encountered geologic conditions. Along most of the tunnel length, the concrete lining is cast in place directly against the rock. This provides for a smooth interior surface. In areas of varying geometry (crossover links, branching points), shotcrete is used for the permanent lining. In areas where water inflows, a waterproof system is incorporated in between the rock support system and the permanent concrete lining. This system includes drainage elements placed against the rock and a waterproofing membrane.

Construction Method

Careful consideration was given to selection of the excavation method early in the project planning phase. The Lötschberg base tunnel was split into sections that can be optimally constructed in technical terms. The Mitholz, Ferden, and Steg access tunnels serve as intermediate points of attack. From their base points, as well from the portals, the excavation of 11 tunnel sections can be embarked upon. These sections were planned with an optimal use of TBMs and, at the same time, a minimum risk of delays. From the south, two tunnel boring machines are carving their way through the rock: for the Steg lateral adit/base tunnel and for the first 10 km of the eastern tube from the south portal

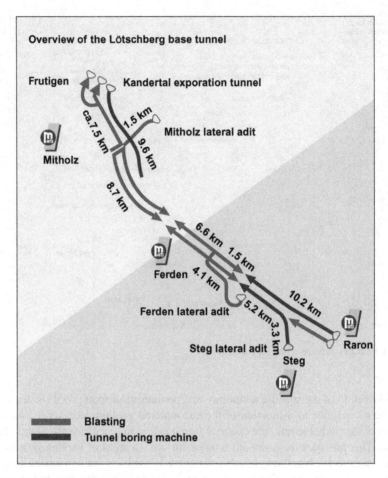

FIGURE 8.26   Overview of the Lötschberg base tunnel.

at Raron. In all the other sections, the traditional drill-and-blast method is being used (Fig. 8.26).

Difficulty in Geology of Lötschberg Base Tunnel

The two "sedimentary slice" and "Jungfrau Wedge" intrusion zones were investigated starting from the foot at Ferden in a northerly direction. These zones had to be crossed by the following three tunnels: (a) the western base tunnel in the northern direction, with an excavation area of about 65 m², (b) the eastern base tunnel in the northern direction, and (c) the entry/ventilation adit, with an excavation area of about 40 m².

The depth of the Aar massif granite and granodiorite rock cover along the Lötschberg base tunnel route reaches up to 2000 m in two places. The depth is greater than 1500 m for a total of about 9.3 km (Fig. 8.27).

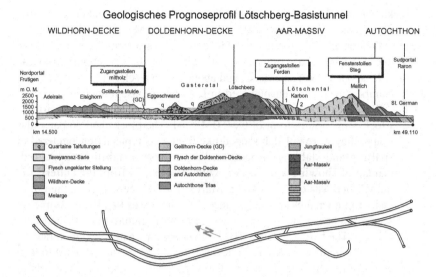

FIGURE 8.27 Geological longitudinal profile for the LBT.

Experience of Lötschberg Base Tunnel Excavation

Rockburst

- Theoretical analysis
  For the evaluation of zones with rockburst risk and their location with high overburden, the engineering community IGWS applied the following procedure in the first phase:
  - An investigation of the natural geostatic stress state of rock mass on the basis of a three-dimensional finite element model on the assumption that the controlling stress state was determined by the relief and not by tectonics
  - Comparison of modeling results with estimated stresses from hydrofracturing tests in boreholes
  - Back analysis of rockburst phenomena in the existing Lötschberg apex tunnel
  - Evaluation of the risk zones in the Lötschberg apex tunnel on the basis of values determined in the apex tunnel under the rockburst critical point calculated from (c).
- Empirical analysis
  From experience with the Lötschberg apex tunnel and from comparisons with other underground construction with rockburst problems (Mont Blanc tunnel), the following classification was formulated: class A = very high rockburst risk (75 to 100%) with $\sigma_1 > 130$ MPa; class B = high rockburst risk (50 to 75%) with 120 MPa $< \sigma_1 <$ 130 MPa; class C = medium

rockburst risk (25 to 50%) with 110 MPa $< \sigma_1 <$ 120 MPa; and class D = low rockburst risk (0 to 25%) with 100 MPa $< \sigma_1 <$ 110 MPa. On the basis of this classification and the aforementioned theoretical considerations, about 4.1 km of the Lötschberg base tunnel were classified as class A, about 1.4 km as class B, about 1.4 km as class C, and about 300 m as class D.

- Rockburst phenomena encountered in Lötschberg base tunnel
  For TBM excavation, two distinct failure phenomena could be identified:
  - Block formation in front of the TBM head: The blocks had no particular shape; they were not slab shaped like fragments typically encountered in spalling and slabbing processes. When these instabilities appeared, instead of showing a flat aspect with clear marks from the cutters, the tunnel face was quite irregular. Marks from the cutters were then only visible in a smaller area of the face. In some cases this block formation seemed to be assisted by the presence of weak failure planes filled with chlorite. The blocks reduced TBM utilization time and penetration rate.
  - Onion skinning initiated at the level of the TBM shield: About 0 to 4 m behind the tunnel face, scales of low thickness peeled off the walls in the excavation. When high overburden depth was encountered, deep notches up to about 1 m may appear, typically in a symmetrical pattern. In some cases, notches prevented grippers from making good contact with the rock mass, which caused significant output losses. Most often, these phenomena were noticed in zones of strong massive rocks. Most strain energy releases occur within 4 m from the face within the TBM shield. There was no evidence that the TBM cannot handle the minor strain energy releases experienced so far.
- Practical measures for reducing effects of rockburst

The TBM excavation in the areas of rockburst was manageable with standard, good quality tunneling procedures, as there was no evidence of violent strain energy release. The rock support consisted of Swellex expansion bolts and steel nets, fiber shotcrete before bolting, spray-on linings, etc. The 4.0-m-long TBM shield provided protection for workers and equipment but was a handicap in controlling the bulking process in the first few meters behind the face.

For tunnel blasting, a special profile type was designed in order to control rockburst: following the immediate protection with a shotcrete layer reinforced with steel fiber, a double-twisted wire netting is attached. In addition to this wire netting, 3.0- to 3.5-m-long expansion bolt anchors with a deformation capacity of 35 mm are mounted in a grid measuring 1.5 × 1.5 or 1.0 × 1.0 m. Protection plates with dimensions of 22 × 50 cm provide additional protection. In the case of violent rockburst phenomena, the tunnel header is separated into a calotte and bench. Further measures include shortening the bore stroke, relief blasting in the region of the tunnel walls, and possibly changing and detonating the systematic preliminary boreholes.

Rock mass sealing of a high-pressure, water-conducting zone
for environmental reasons

The two "sedimentary slice" and "Jungfrau Wedge" intrusion zones were investigated, starting from the foot at Ferden in a northerly direction. Mountain water in the intrusion zones had a temperature up to 39.5°C and a fairly high sulfate content, up to 2 g/liter. This placed certain requirements on the injection material used and on the protection of extension of the tunnels. In these cases, the following procedure was laid down as a basic principle: only that which was absolutely necessary would be carried out with the absolute minimum of material, but upgradeable in order to subsequently attain the required results. This meant that the drilling was designed and executed so that at the same time it could be used for checking, drainage, and injection drilling. In the case of injection, this was a sealing injection. The target was therefore to make a leak-proof ring at some distance from the tunnel.

REFERENCES

[1] Sharma VM. Metro stations and rail tunnels of metro projects. In: Proceedings national workshop on underground space utilisation. New Delhi: ISRMTT; 1998. p. 131–63.

[2] ITA. Examples of benefits of underground urban public transportation systems, ITA working group on costs-benefits of underground urban public transportation. Tunnel Undergr Space Technol 1987;2(1):5–54.

[3] Heijboer J, van den Hoonaard J, Van de Linde FWJ. The Westerschelde tunnel: approaching limits. Gorter, Steenwijk, The Netherlands: A. A. Balkema Publishers (Swets & Zeitlinger); 2004. p. 292.

[4] ITA. Underground or above ground? Making the choice for urban mass transit systems. Tunnel Undergr Space Technol 2004;19:3–28.

[5] Bhasin R. Personal communications with Bhawani Singh at IIT Roorkee. India; 2004.

[6] Kaiser PK, McCreath DR. Rock mechanics considerations for drilled or bored excavations in hard rock. Tunnel Undergr Space Technol 1994;9(4):425–37.

[7] Bickel JO, Kuesel TR. Tunnel engineering hand book. New York: Van Nostrand Reinhold Company; 1982. p. 670.

[8] Gulhati SK, Dutta M. Geotechnical engineering, chapter 28. New Delhi: Tata McGraw Hill Publishing Co. Ltd.; 2005. p. 738.

[9] Singh M. Delhi metro line No. 2—tunnelling by shield tunnel boring machines. In: Proceedings seminar on productivity and speed in tunnelling. CBIP, Dehradun, India; 2003. p. 207–18.

[10] Broomfield JD, Denman D. TBM rock tunnelling in India, seminar on productivity and speed in tunnelling. CBIP, Dehradun, India; 2003. p. 194–206.

[11] Dodeja SK, Mishra AK, Virmani RG. Successful utilization of double shielded TBM in low cover zone for excavation of inclined pressure shaft in Parbati H.E. Project II (NHPC). In: Sharma KG, Mathur GN, Gupta AC, editors. Workshop on rock mechanics and tunnelling techniques. Sikkim, India: ISRM (India) and CBIP; 2007. p. 169–84.

[12] Bieniawski ZT, Celeda B, Galera JM. Predicting TBM excavability. Tunnels and tunnelling international, Progressive Media Publisher, Kent, UK, September; 2007. p. 4.

[13] Singh B, Goel RK. Engineering rock mass classification: tunnelling, foundations and landslides. Elsevier Inc., USA; 2011, p 365.

[14] Vandebrouck P. The channel tunnel: the dream becomes reality. Tunnel Undergr Space Technol 1995;10(1):17–21.

[15] Agarwal R, Gupta AK. Rock excavation by controlled blasting in underground metro corridor of Delhi MRTS. Proceedings ISRM symposium—advancing rock mechanics frontiers to meet the challenges of 21st century. CBIP, New Delhi; 2002. p. IV.67–IV.78.

[16] Anbalagan R, Singh, Bhawani, Bhargava P. Half tunnels along hills roads of Himalaya: an innovative approach. Tunnel Undergr Space Technol 2003;18:411–9.

[17] Amundsen FH, Hord A. Most accidents at tunnel entrances. Norwegian road tunnelling. University of Trondheim, Norway: Tapier Publishers, publication No. 4; 1982. p. 27–32.

[18] Lunderbrekke E. Tunnels as elements of the road system. Norwegian road tunnelling. University of Trondheim, Norway: Tapier Publishers, publication No. 4; 1982. p. 9–18.

[19] Barton N. Lecture on tunnel support selection from Q classification and support element properties, lecture No. 3, training course on rock engineering for large hydropower projects. New Delhi, India: CSMRS; 2011a.

[20] Hoek E. Big tunnels in bad rocks, 37th ASCE, Karl Terzaghi lecture. J Geotechn Geoenvironment Eng 2001;127(9):725–40.

[21] Barton N. An engineering assessment of pre-injection in tunnelling. J Rock Mechanics Tunnel Technol 2011b;17(2):65–81.

[22] Henning JE. Longitudinal ventilation of road runnels. Norwegian road tunneling. University of Trondheim, Norway: Tapier Publications, publication No. 4; 1982. p. 133–46.

[23] Barton N. Q_{slope} method, training course on rock engineering. New Delhi: CSMRS; 2008. p. 343–63.

[24] Singh B, Goel RK. Software for engineering control of landslide and tunnelling hazards. Krips, Meppel, The Netherlands: A.A. Balkema (Swets & Zeitlinger); 2002. p. 344.

[25] Inokuma A, Inano S. Road tunnels in Japan: deterioration and countermeasures. Tunnel Undergr Space Technol 1996;11(3):305–9.

[26] Marec M. Major road tunnel projects, how far can we go? Tunnel Undergr Space Technol 1996;11(1):212–6.

[27] Martin D. Vardo tunnel—an undersea unlined road tunnel Norwegian style. Norwegian hard rock tunnelling. University of Trondheim, Norway: Tapir Publishers, publication No. 1; 1982. p. 100–2.

[28] Blindheim OT, Grov E, Nilsen B. Norwegian sub-sea tunnelling. Norwegian Tunnel Soc, Sandvika, Norway, Publication No. 15; 2006. p. 89–97.

[29] Culverwell DR. Comprehensive merits of steel and concrete forms of tunnels. In: Proceedings conference immersed tunnel techniques. Manchester: Institution of Civil Engineers; 1990. p. 185–98.

[30] Asakura T, Kojima Y. Report to ITA working group on maintenance and repair of underground structures, state-of-the-art of non-destructive testing methods for determining the state of a tunnel lining. Tunnel Undergr Space Technol 2003;18:161–9.

[31] Haack A, Schreyer J, Jackel G. Inspection, maintenance and repair of tunnels: international lessons and practice. Tunnel Undergr Space Technol 1995;10:413–26.

[32] Richards JA. Tunnel maintenance in Japan. Tunnel Undergr Space Technol 1998;13:369–75.

[33] Ikuma M. Maintenance of the undersea section of the Seikan tunnel. Tunnel Undergr Space Technol 2005;20:143–9.

Chapter 9

Underground Storage of Crude Oil, Liquefied Petroleum Gas, and Natural Gas

Mother Nature often compensated for engineering deficiencies.

JC Rogiers [1].

9.1 INTRODUCTION

The urge to store oil was first felt by oil-starved countries that depended on imports. The Scandinavian countries took the lead in this direction by building a number of underground storages. The sudden oil price rise in the 1970s compelled many countries to build underground storages of oil. Such countries prepared plans and constructed storage in a phased manner. During the last four decades, underground storage has grown considerably. Old tankers, pipe lines, abandoned mines, salt domes, and solution cavities were evaluated for possible storages, and some countries succeeded in building requisite capacities. Freshly excavated caverns were also built in promising areas.

Broadly, the storage technique can be grouped into three classes: (a) above-ground storage, (b) inground storage, and (c) underground storage (Fig. 9.1). This chapter presents technological aspects of the underground storage of oil, liquefied petroleum gas (LPG), and other petroleum products.

Before discussing the underground storage technology for oil, LPG, and other petroleum products, investigations generally carried out for such storage facilities are listed briefly.

9.2 INVESTIGATIONS AND DESIGN

For a cavern project, a typical site investigation follows these steps:

1. Site selection and geomapping
2. Seismic surveying
3. Core drillings, water loss measurements, and water table observations
4. Logging of core samples and rock mass quality assessment
5. Laboratory tests
6. Inspection, monitoring, and control during excavation

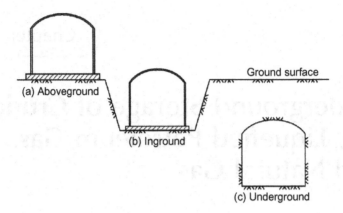

FIGURE 9.1   Various forms of oil storage prevail the world over.

Selmer-Olson and Broch [2] summarized the design of cavern projects as consisting of four distinct steps.

1. *Selection of location:* This is the most important single step in the design chain, and every effort should be directed toward making the right selection.
2. *Orienting the longitudinal axis of the opening:* This is to be done so as to minimize stability problems and overbreak (see Section 4.4).
3. *Shaping the opening:* Properly shaped openings will favorably distribute stresses along the periphery.
4. *Dimensioning the opening:* In Norway, this is usually based on empirical techniques. Economic and operational considerations must also be taken into account here.

9.3 UNDERGROUND STORAGE TECHNOLOGY

The following types of underground storage technologies are in application:

- Underground unlined storage
- Underground lined storage

In addition, a partly underground and partly on ground facility composed of (a) steel and concrete structures built by the cut-and-cover method and (b) steel tanks buried or sheltered in earth is also being used.

9.3.1 Underground Unlined Storage

More than 200 facilities for the underground storage of petroleum products in unlined caverns have been completed successfully in Finland, Norway, and Sweden since the mid-1960s.

For successful application of this storage technique, four prerequisites must be fulfilled [3]:

1. Competent and stable rock conditions, suitable for construction of large openings, must exist (see Section 4.4).
2. The stored product must be lighter than water.
3. The stored product must be insoluble in water.
4. The rock surface in contact with the stored product must hold a pore water pressure higher than the static pressure exerted by the stored product.

The principle applied for storage varies depending on the product to be stored. Products may be stored either at atmospheric pressure or at high pressure.

Storage under Atmospheric Pressure

Products such as kerosene, gasoline, light oil, and heavy fuel oils are usually stored under atmospheric pressure. This is carried out in the following way (Fig. 9.2).

1. A cavern is blasted out immediately below the absolute lowest groundwater level.
2. Groundwater leaking into the cavern is pumped out, with the result that the groundwater table around the cavern sinks until it is on a level with the bottom of the cavern. Continuous pumping ensures that the groundwater surface is maintained at this lower level.
3. The product can now be stored in the cavern, floating freely on the groundwater surface at the bottom of the cavern.

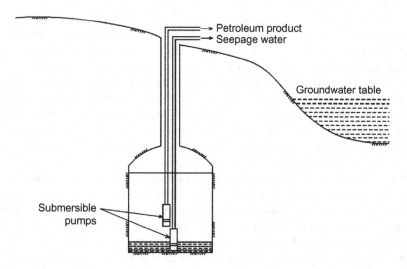

FIGURE 9.2　Petroleum product stored in a cavern under atmospheric pressure.

Note: There are generally two regulations applied to the unlined underground storage of petroleum products: (i) the maximum height of the product surface must be at least 5 m below the lowest groundwater surface in the surrounding area and (ii) there must be space in the cavern for water leaking in for at least 3 days.

These caverns are completely separated from the ground surface. All water in the caverns originates from groundwater seeping in through the cavern walls. This inflow occurs during operation at low gas pressure. The water is collected in a sump excavated in the cavern floor. From the sump, the water is displaced to the surface by use of gas pressure and by a submersible electric pump (Fig. 9.2).

Thus, essentially all cavern space is used for storage. When gas is injected into the cavern, the pressure increases gradually. The maximum allowable pressure is near the hydraulic pressure of the pore system of the rock surrounding the caverns.

Normally, the depth of the cavern will dictate the hydrostatic pressure. However, by using an artificial water injection system (water curtain), a gas pressure higher than hydrostatic may be allowed, thereby improving the economics of storage significantly.

Water Curtain

Physically speaking, gas leakage from an unlined rock cavern is possible only along rock joints that develop declining hydrodynamic pressure away from the cavern as the gas pressure is raised. Thus, any occurrence of leaky rock joints may be prevented by introducing a controllable outer boundary condition for the groundwater pressure. A water curtain is a system of drill holes used to achieve this leakage control.

The need for a water curtain around gas storage caverns is particularly evident if storage pressures above the normal hydrostatic groundwater pressure are to be used. Depending on the degree of overpressurization and the rock joint geometry, the water curtain may attain one of two forms, roof curtain or roof-and-wall curtain. These are depicted in Figure 9.3. A water curtain consists of a series of parallel holes drilled from a small tunnel (drift) by percussion drilling tools. After completion, each hole is connected to a water charge system running inside the tunnel. Spacing and direction of the drill holes are determined on the basis of the geometry and hydraulic characteristics of the rock joint system. After all of the curtain drill holes are connected, the system is operated from a control station at the surface. Watertight plugs are constructed at the ends of the tunnel. The main water line is then led through a vertical drill hole to the water tunnel. The water tunnel itself will be filled with water through the same hole and then pressurized to the same level as the curtain drill holes. In addition to the primary purpose of hydrodynamic pressure control, another important function of the curtain is to ensure full water saturation in the rock joint system at all times. Operational experience with water curtains used for storing liquid

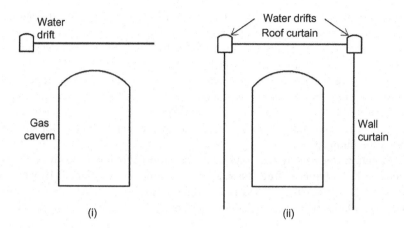

FIGURE 9.3   Water curtain configurations: (i) roof curtain and (ii) roof and wall curtain [4].

petroleum and air has demonstrated that recharging of the joint system or rock masses is possible, should invasion of gas or air occur accidentally.

The water curtain should be operated at a minimum pressure differential to the gas pressure. At this pressure differential, the water pressure in the most critical rock joint will just exceed the gas pressure and water flow will be at a minimum. However, all other joints connected hydraulically with the curtain will discharge some water to the cavern. The critical pressure differential is a complex function of the geometry and hydraulic characteristics of the rock joint system and may be determined only approximately by calculations. Optimal conditions must be determined through test operation of the cavern with air after construction is completed.

During periods when the gas storage is operated at pressures higher than the ambient groundwater pressure, there will be an unavoidable discharge of injected water away from both cavern and water curtain installation into the regional groundwater regime.

An unlined gas cavern with a water curtain is characterized by quite a high degree of reliability due to the following main factors:

- Site- and system-specific, safe operational conditions may be outlined at the outset of operations.
- Monitoring of gas leaks is performed continuously in the rock mass.
- Any accidental gas invasion may be "repaired" by engaging an elevated water curtain pressure until normal hydrodynamic conditions are reintroduced.

Concept of Fixed and Mobile Water Bed

While understanding the storage method it is very essential to know the concept of a fixed and mobile water bed. As mentioned earlier for storage at atmospheric pressure, the water bed is brought down by pumping and is maintained at a constant level near the cavern base. The petroleum product being stored floats on the water bed; the level of its upper surface is controlled by a product pump. Because the water level at the base is constantly maintained, this procedure is known as storage on a fixed water bed.

In a mobile water bed system, the cavern is filled with the petroleum product up to the ceiling height and is maintained at this level by pumping out water. Thus the level of water fluctuates in accordance with the quantity of stored product. If more quantity is to be stored, an equal amount of water for storing should be pumped out.

Mobile water bed storage is favored for more volatile petroleum products because the volume of free space above the product is reduced. However, fixed water bed storage is used more commonly when storing heavy crudes and less volatile fuel oil as it requires less pumping and adjustments of fluid levels.

Storage under High Pressure

In case of pressurized storage, the rock cavern is located at such a depth below the groundwater level that the hydrostatic pressure exceeds the gas liquefaction pressure at the temperature of the rock. This means, for example, that a storage plant for propane must be located approximately 100 m below the groundwater level to meet the pressure requirements (requisite pressure approximately 9.0 bar at a rock temperature of approximately 25°C). The principle has been used successfully in Sweden and other European countries (Figs. 9.4 and 9.5).

Refrigerated Storage

For refrigerated storage, the rock cavern is kept at a gas liquefaction temperature, which for the most common petroleum gases, propane and butane, is approximately −40° and −20°C, respectively, at atmospheric pressure. Such installations again exist in Sweden. Both plant and operating costs have proved to be advantageous compared with conventional storage above ground. This is based on the principle of refrigerating the rock around the cavern so that all groundwater in the fractures freezes. In this way an impermeable zone of ice (gas-tight envelope) is formed all around the cavern (Fig. 9.6). The frozen zone, thus created, grows up continuously and reaches up to about 30 m in 30 years. The rock cavern, therefore, should be located more than 30 m below the groundwater table.

Underground storage of LPG can be carried out in two ways: storage under high pressure and refrigerated storage described previously.

The storage of LPG or liquefied natural gas (LNG) under very low temperature conditions is highly desirable. In this context, properties of rock at very low temperatures should be obtained to know the behavior of rock mass and the stability of the underground opening under refrigerated conditions. Further,

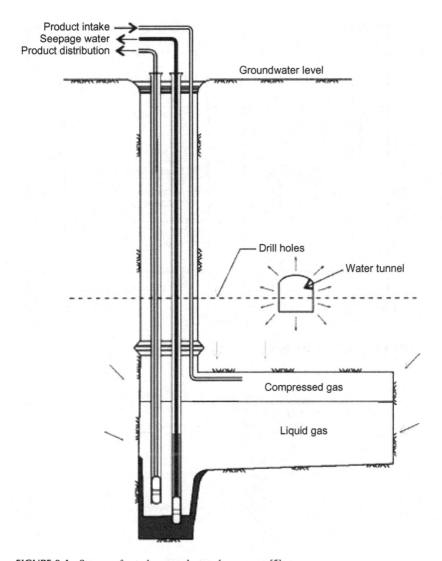

Product intake →
Seepage water ←
Product distribution ←

Groundwater level

Drill holes

Water tunnel

Compressed gas

Liquid gas

FIGURE 9.4 Storage of petroleum product under pressure [5].

rock properties and behavior are also studied under cyclic freezing-and-thawing conditions (see Fig. 9.7).

Some technical problems remain to be solved in designing and constructing underground storage plants under refrigerated conditions. For example, heat supplied from the groundwater flow crossing from the outer boundary some-times interrupts formation of the frozen zone around the cavern. This occurs especially in cases where the flow rate or velocity is too high. In other cases,

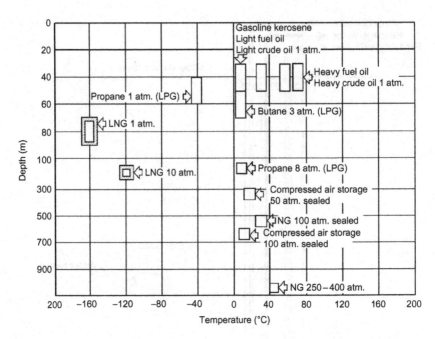

FIGURE 9.5   Temperature and depth requirements for various forms of energy storages [6].

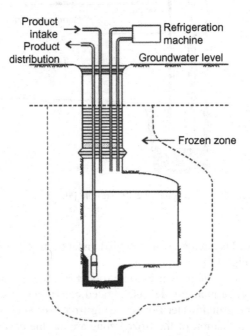

FIGURE 9.6   Sketch showing principles of refrigerated storage.

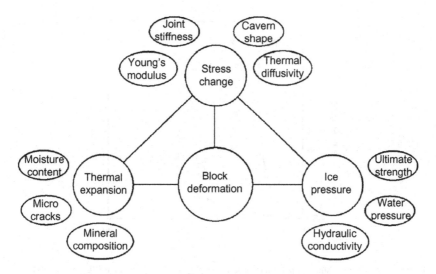

FIGURE 9.7 Factors and rock parameters that affect thermal deformation of a block in the rock mass surrounding a refrigerated cavern [7].

support systems such as rock bolts and grouting material may suffer from an unacceptably low temperature condition [8].

To solve these issues, generally, bench-scale tests are conducted to determine the critical water flow and velocity, which relate to interruption of the creation of a frozen zone.

9.3.2 Underground Lined Storage

As the name indicates, this is a system of storage in which the stored product remains in contact with the lining applied all around the exposed rock of the cavern. The lining could be plastic or steel. Underground lined storage is contemplated in preference to other methods of storage when the site is unfit due to bad-quality rock or in such regions where no groundwater table is present or where an artificial water table cannot be established. The storage cavern can be excavated at a shallow depth or in a hillock. Access to the cavern is by means of a drift or tunnel that is usually very short in length compared with an unlined cavern. Such storage can be operated at very high pressures even at moderate depths. Although lining in combination with rock presents an extremely stable design to withstand internal pressure, under empty conditions the hydrostatic pressure may act from outside, which may pose a threat to storage. It is therefore necessary to provide proper drainage around the cavern to make it safe.

Lined caverns are most suited for refrigerated storages. They are very efficient and the power for maintenance may be reduced by 50 to 60% of that

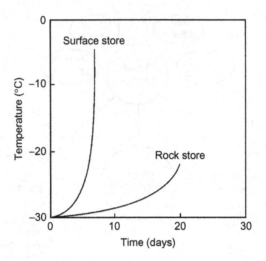

FIGURE 9.8   Temperature increase during stoppages in refrigerated storage [9].

needed for surface storage. In the context of developing countries where frequent power breakdowns are common, such stores are very useful. Figure 9.8 shows the rise in temperature due to power failure with time. It may be seen that a surface storage tank loses refrigeration very fast, that is, within 5 days, whereas underground storage takes 20 days or more.

Petrol, diesel, spirits, and gas can be stored without fear of any loss due to leakage. Breakdown of the water pressure mechanism may lead to leakage in unlined caverns. This possibility is completely ruled out in lined caverns. Lined caverns can be pressurized to the desired extent and may thus be most suited for LPG and natural gas. For special products, such as jet fuel, it is important to prevent any contact with water. In this case, storage over a water bed, therefore, is avoided. Moreover, developed nations should have underground storage of crude oil of at least 3 months to meet demands of wartime.

The whole cavern is lined and sealed with shotcrete. These linings can be made strong enough to fulfill the extremely high requirements of chemical resistance and to seal cracks, even those with great variations in width. A simple and comparatively inexpensive method for sealing concrete and rock has been used in Sweden since 1958. The seal consists of plastic layers on a concrete base or on smooth gunite/shotcrete in a rock cavern.

Advantage

Underground lined storage has the following advantages:

- Nonflammability.
- Strong resistance to external water pressure.
- Strong resistance to cracking in supporting wall.

- Easier accessibility.
- It can be planned even in a comparatively bad or weak quality of rock.

An unlined cavern disturbs the groundwater regime. The groundwater may get contaminated with oil. An emulsion may form, which can damage the aquifer permanently. Moreover, disturbance of the groundwater regime may also be a cause of concern.

Table 9.1 gives types of storages and appropriate storage materials for underground storage techniques.

Lined caverns are favored because of the following reasons.

- Caverns excavated in competent rock formations can be made leak proof at an additional cost of 15 to 20% of an unlined cavern's cost. This extra cost is insurance against possible leakage of material in case of unlined caverns. The advantage of lined caverns may outstrip the high initial cost over a period of time.
- Natural groundwater pressures are created by artificial means to contain the oil. Ground water pressure depends on the groundwater table and the porosity of the rock medium, as these may change with time and thus may not be in conformity with the theoretical design stipulations. Such a situation may, therefore, lead to leakage. This problem is avoided altogether in lined caverns.
- In case of unlined caverns, the water comes into contact with the crude oil and products. Crude oil had been found to be unaffected by the water however, micro biological activities at the interface of water and the stored products had been found to affect its quality adversely.
- Unlined caverns disturb the groundwater regime greatly; in case of arid and semiarid conditions, the precious fresh water meant for human consumption is lost.
- The groundwater thus gets contaminated and sometimes becomes unfit even for domestic and industrial use. Underground storage facilities, either lined or unlined, consist mainly of two types of underground structures. These are described in the following section.

9.4 STORAGE OF NATURAL GAS

Natural gas (NG) is the most difficult gas to store. As the use of natural gas has increased considerably, the need to build storage facilities for it has grown accordingly. Natural gas is liquefied using high pressure and a low temperature (Fig. 9.5). Storing NG in surface storages is expensive because the temperature of the gas has to be reduced to below −160°C in order to liquefy it under normal atmospheric pressure. In rock storages, however, groundwater pressure can be utilized because the gas liquefies at a higher temperature when it is under higher pressure. Rock storage of NG is currently being studied in a number of countries, including Finland, Norway, and Sweden [9].

TABLE 9.1 Product Wise Suitability for Underground Storage [10]

Techniques	Gas: Natural Gas/ Compressed Air	Liquified Under Pressure Gas: Butane/Propane/ Propylene Ethane/ Ethylene	Liquified at Low Temperature Gas: Propane (–40°C)/ Ethylene (–105°C)/ Natural Gas (–162°C)	Liquids: Crude Oils/ Petroleum Products/ Hot Water	Wastes: Industrial/ Nuclear
			Products		
Aquifers	1	2	2	2	2
Leached cavities	1	1	2	1	1
Mined caverns	1	1	2	1	1
Disused mines	1	1	2	1	1
Cryogenic caverns	2	2	1	2	2

1 - Usable technique
2 - Non usable technique

In 1961, an abandoned coal mine (Leyden) near Denver, Colorado, was converted into natural gas storage at a 250- to 300-m depth. Overlying the mine are permeable sandstone and impermeable schists. The maximum allowable gas pressure was determined at about 20 bar (2 MPa) and the minimum pressure at 4 bar (0.4MPa), giving a total storage capacity of about 85 MNm^3 (2300 mscf) of gas. Only about half of the injected gas is available on a short-term basis, as much of the gas dissolves in the surrounding coal formations. Minimum depletion time is 5 days. This storage has been of considerable aid to the city of Denver in the management of its greatly variable gas load in the wintertime.

Two coal mines in Belgium and one in Germany operate with natural gas storage under conditions similar to those of the Leyden coal mine.

Shale is often impregnated with oil and natural gas. New technology has advanced. A shale gas revolution has converted the United States from a gas importer to a gas exporter. As a result, American gas prices have fallen rapidly to as low as US$12 per barrel. Europeans see gas as a means to break dependence on Russian gas. Shale gas reduces carbon emissions significantly. India is also rich in shale gas. China is also using horizontal drilling technology and hydrofracturing with water and sand to extract shale gas.

9.5 TUNNEL-SHAPED STORAGE FACILITY

Storage caverns have usually been constructed horizontally with tunnel (arch)-shaped caverns (Fig. 9.9) with a span up to 20 m, depth to 35 m, and length up to several hundred meters [12]. A tunnel shape probably was

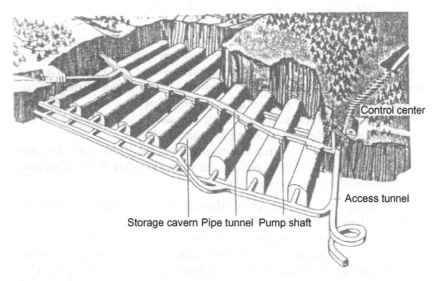

FIGURE 9.9   Horizontal tunnel-(arch) shaped cavern for oil storage [11].

chosen as a consequence of the development of efficient tunnel excavation equipment during the last decades. This design has proven to give safe and inexpensive operation of storage. Tunnel-shaped caverns have the following disadvantages.

9.5.1 Disadvantages

Experience shows that storage on a large extended water bed is not suitable for some petroleum types, for example, heavy crude types prone to create sludge deposit problems. Also, a large water bed is unsuitable for storing products that are sensitive to water such as jet fuels. In fact, jet fuel can be affected adversely by bacteria and fungi to such an extent as to be disastrous for safety in flights [12].

9.5.2 Separation Distance between Caverns

The separation distance between two caverns is determined from mechanical conditions and hydraulic conditions and must satisfy both. The objective of the former is to relieve mechanical interaction between caverns; the objective of the latter is to prevent leakage of the stored oil between caverns.

According to the aforementioned official permit requirements, mechanical conditions are determined unconditionally by the formula given here. With respect to hydraulic conditions, safety against oil leakage is to be confirmed by performing a finite element flow analysis, assuming the state characterized by the most severe conditions, that is, the case in which a filled cavern and an empty cavern are located adjacent to each other [13].

$$L = [(B1 + H1 + B2 + H2)/4] + R1 + R2, \tag{9.1}$$

where L is minimum distance to be secured (distance between walls of adjacent caverns), B1 is maximum internal width of one cavern, H1 is maximum internal height of the same cavern, B2 is maximum internal width of the adjacent cavern, H2 is maximum internal height of the adjacent cavern, R1 is width of loosened zone in rock mass surrounding one cavern, and R2 is width of loosened zone in rock mass surrounding the adjacent cavern.

The evaluation of loosening and estimation of the width of the loosened zone may be carried out by finite element method (FEM). Further, supports may be installed in the following sequence:

- First of all, a 50-mm thickness of shotcrete/steel fiber-reinforced shotcrete (SFRS) may be sprayed to stabilize the cavern temporarily.
- Then a rock bolt system may be installed according to the design.
- Finally, the remaining layers of shotcrete/SFRS may be sprayed according to the design thickness or until instruments indicate that a cavern has become stable.

9.6 MULTITANK STORAGE (POLYTANK) CONCEPT

In some situations, however, tunnels or large rooms with horizontal axes are less suitable. Construction in soft rock often requires expensive reinforcement of rock. In areas with limited underground space it is favorable to extend the storage downward in a vertical direction; the tank thus created is called a polytank. The polytank system is very versatile as it can be used to solve design and construction problems. It will also improve and simplify the maintenance and operation of the storage.

9.6.1 Construction Principles

Polytank storage consists of a group of cylindrical caverns with vertical axes (Fig. 9.10). The shapes of the cross section of the caverns and the arrangement of the caverns are dependent on the in situ state of stress and the rock mass quality. For the storage of oil, the group is favorably designed with a central shaft, where pumps and most of the control devices are placed. The storage caverns are placed in one or several hexagons or rings around the shaft. They can be given the same shape and size and will preferably be placed on the same level below the groundwater table. The minimum pillar width should be adequate for its stability.

As the roof of an upright cylindrical cavern is small compared to the roof of a conventional horizontal storage tunnel of the same capacity, the risk of rock fall from the roof is less because of more stability.

If different products are to be stored in separate caverns, the caverns can be separated from each other by concrete barriers in the connecting tunnels. The product can be pumped with a separate pump in pump shafts for each cavern.

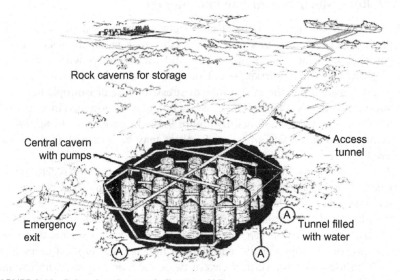

FIGURE 9.10   Polytank underground oil storage [14].

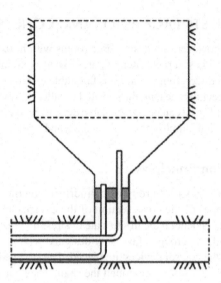

FIGURE 9.11   Cavern bottom and funnel design [14].

From the point of view of operation and maintenance of the cavern, the bottom is funnel shaped, as shown in Figure 9.11. The bottom funnel with a short vertical shaft down to a horizontal tunnel grid leads to pumps in the central cavern or shaft.

9.6.2 Rock Mechanics of Polytank Storage

The size of the chambers and the distance between them are dependent on the quality of the rock mass. In a very good quality rock mass, chambers can be constructed with the same volumes as valid for chambers with horizontal axes. However, for the arrangement in space of the chambers one has to take into consideration the in situ state of stress. In case of isotropic horizontal stress, chambers with a circular cross section may be arranged in a circular or hexagonal pattern, as shown in Figure 9.12. In case of a strong anisotropy of horizontal in situ stresses, chambers may be arranged as shown in Figure 9.13.

FEM studies show that a vertical design is very favorable with regard to distribution of rock stresses around the cavern or cavern group. It is possible to avoid tension areas in the rock walls, and their compression stresses can be kept within moderate limits [15]. This implies, in brief, that within a given rock mass, vertical caverns can be excavated to a larger cross section than horizontal caverns. Comprehensive calculations indicate that in most cases it is possible to use spans for vertical cavern twice (or more) those of horizontal ones [15].

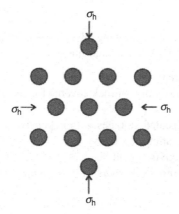

FIGURE 9.12 Arrangement of multitank storage (polytank) caverns where horizontal stresses are the same in magnitude [14].

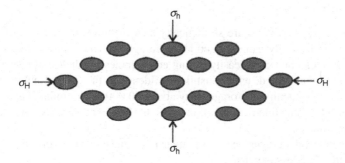

FIGURE 9.13 Arrangement of multitank storage (polytank) caverns where horizontal stress (σ_H) is much greater than other horizontal stress (σ_h) [14].

For underground storage in rock mass of poor quality, particularly in horizontal and layered rocks, the polytank concept offers considerable advantages.

9.6.3 Field of Applications

There are the following field applications:

- Oil storage on water bed in unlined caverns
- Oil storage in lined (sealed) caverns
- LPG storage on a water bed in unlined, pressurized caverns
- LPG storage in sealed and insulated caverns
- LNG storage in sealed and insulated caverns
- Other miscellaneous applications

9.6.4 Advantages

The polytank concept has the following advantages in comparison with conventional rock storage with large horizontal tunnels:

- Better use of land area
- Better durability and stability of rock caverns for certain rock types and rock qualities
- Better economy for sealed storages with expensive lining
- Better and safer operation of storage
- Less impact on the environment
- Better thermal economy for insulated caverns (LNG and hot water storages)

9.7 GENERAL ADVANTAGES AND DISADVANTAGES OF UNDERGROUND STORAGES

9.7.1 Advantages

- Underground storages are strategically safe from a warfare point of view. They are naturally camouflaged and safe against small bomb attacks.
- Storages can be tapped easily during emergencies and hence domestic supplies are maintained, which instills confidence in the general public.
- For a large quantity of storages (say above 50,000 m^3 for unlined and above 200,000 m^3 for lined caverns), caverns are less expensive as compared to surface tanks.
- Underground storages need no regular maintenance. Therefore, in the long run they are less expensive.
- Because underground storages need little area of land aboveground, there is not much cost involved for land.
- In view of improved safety, the insurance cost is very low.
- The excavated rock may have some economic value as it may be used as building material.
- Stored oil may be sold at higher profit if the oil price escalates suddenly.
- Storages are comparatively safe against fire and sabotage.
- Natural calamities such as flood, typhoons, and earthquakes (<7 M) do not affect underground storages (if more than 10 m below ground and away from faults in good rocks).
- Underground storages in good rocks have a fairly long life as compared to surface tanks.
- The control and loading/unloading can be made simple and cost-effective.
- The natural landscape is not affected as all the installations are underground, hence the scenic beauty of the area is protected. Further, the local ecosystem is not disturbed.

9.7.2 Disadvantages

- Underground storages are less expensive only if good rocks are available.
- There is a risk of oil being lost if leakage takes place.
- Products are affected by microbiological activity if stored in unlined caverns.
- Products lose their potency if stored too long.
- The pumping system sometimes jams and the recovery of oil is jeopardized.
- The oil may affect the groundwater adversely and hence is unsuitable in terrain where groundwater is scarce.
- If not disposed of profitably, excavated muck may cause environmental problems.
- Oil-producing countries discourage sales to countries indulging in storages of large quantities.

  Table 9.1 offers a utility of caverns according to product of storage.

9.8 INGROUND TANKS

Apart from underground storage, oil and its products can also be stored in earth-sheltered or inground concrete tanks. Inground tanks are built by the cut-and-cover method. Many countries have developed inground facilities for product storage in view of various advantages it has over underground facilities. Following are some of the advantages and disadvantages of inground facilities over underground ones.

9.8.1 Advantages

- Concrete tanks are free from corrosion and hence involve lower maintenance costs.
- Tanks are below ground and hence are less affected by temperatures. There are thus very nominal losses due to evaporation.
- In case of refrigerated storages of LNG and LPG, inground tanks are energy efficient. Power consumption is 40 to 50% lower as compared to equivalent tanks aboveground.
- Concrete tanks have a longer life.
- The material of construction is indigenously available, hence no foreign exchange is needed.
- The cost of construction is lower as compared to steel tanks.

9.8.2 Disadvantages

- Concrete tanks have no salvage value and have to be abandoned if not in use. They cannot be dismantled and shifted to another place as may be possible in the case of steel tanks.
- A vapor pressure recovery system is necessary to conserve volatile matter.

Underground storage of oil, LPG, and other petroleum products is popular in developed countries. This technology can also be adopted successfully in developing nations. The need of the hour is to generate confidence among the user industries for such storage. The confidence, in our opinion, can be generated by having at least one such facility for which the strong leader will have to take the lead.

9.9 COST ASPECTS

Numerous cost compilations and comparisons have been made of rock storage caverns and alternative solutions, that is, steel tanks. As a rule of thumb, Norway unlined rock caverns are considered economically favorable for storage volumes of 50,000 m³ (300,000 bbl) and larger.

Over the past two to three decades, innovations in rock-excavating techniques have increased the economic advantages of rock caverns in comparison to steel tanks. This trend is expected to continue.

Cavern cost is also a function of span width. For a Norwegian cavern installation, the relationship shown in Figure 9.14 was found. The discontinuity shown in Figure 9.14 is due to limits for the use of heavy excavation equipment [3].

For a given span, the cost variation in relation to depth can be determined and the solution optimized with respect to cost. In Figure 9.15, each bar is the sum of the following three factors [3].

1. Excavation cost, including drilling, blasting, rock face scaling, mucking, soil transport, and dumping
2. Rock support cost
3. Pumping cost

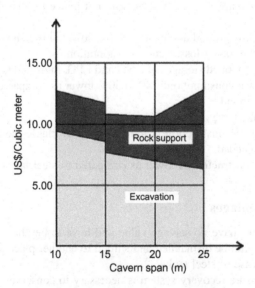

FIGURE 9.14 Excavation and support cost variation with cavern span or width [3].

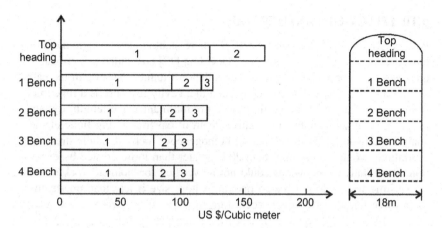

FIGURE 9.15 Cost variation with cavern depth [3].

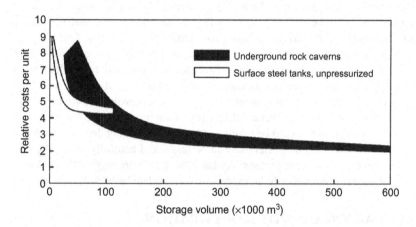

FIGURE 9.16 Relative rock cavern/steel tank costs [3].

The graph (Fig. 9.15) shows that the top heading is expensive, while the average unit excavation cost is reduced after one or two benches are added. Going further down, the gain in excavation cost is lost by the rising rock support cost, increasing pumping costs, and increasing costs of longer shafts and access tunnels and shafts.

Relative cost curves for caverns and steel tanks are given in Figure 9.16 [3]. It may be seen that underground storage is more economical than surface steel tanks for storage volumes of more than 50,000 m³. Israel has developed underground caverns in impermeable chalk for the storage of propane and butane. The cost of storage is about 35% more economical than that on the ground by steel spheres and tanks [16].

9.10 EFFECT OF EARTHQUAKE

Past experience indicates that tunnels are more stable than aboveground structures during an earthquake. However, Dowding [17] concluded that critical frequencies are somewhat lower for caverns than for tunnels because of increased cavern size. Dowding [17] also concluded that cavity response is a function of span or wall height to the earthquake's predominant wave length, dynamic properties of major joints, and relative depth of caverns. Using finite element analysis, Dowding [17] found that for homogeneous rocks, dynamic stress concentrations would be no more than 20% greater than those created by excavation. Quantitative conclusions could not be verified for "jointed" rock masses.

Seismic waves do not see an opening whose size is less than one-tenth of their wave lengths. However, according to [18], seismic rock mass quality ($Q_{seismic}$) is assumed equal to half of rock mass quality Q ($Q_{seismic} = Q/2$).

Yamahara et al. [19] have pointed out that the intensity of earthquake motion at deep base rock may be from one-fourth to one-third of the intensity on the ground surface. They also concluded that rock caverns are extremely safe from earthquakes, given that the rock is practically homogeneous, if the cavern is more than 10 m below the ground surface as Rayleigh waves may not be deeper than that and the cavern is away from faults in good rocks.

The damage of supports of underground openings is insignificant from earthquakes of intensity ≤7 M, irrespective of distance from an active fault. Major earthquakes cause major damage to supports near plastic fault zones, which are repairable. The plain cement concrete (PCC) lining of tunnels needs to be grouted well with the rock mass so that bending stresses are minimal. The innermost free surface of PCC lining should be reinforced to prevent spalling due to reflection of seismic waves [20]. The portals should be stabilized carefully with long rock anchors in highly seismic regions. Aydan et al. [20] also suggest that enlarged underground openings should be made to accommodate large deformations.

9.11 CARBON DIOXIDE SEQUESTRATION

Broadly, carbon dioxide (CO_2) sequestration is any means that prevents carbon dioxide from entering the atmosphere. Realistically, this is equivalent to the permanent storage of carbon dioxide, independent of the state of the CO_2 or the location of the storage. Traditionally, carbon sequestration has referred to the carbon stored in biomass, but the definition has come to include the disposal of carbon dioxide underground or in the ocean, as well.

It is known that increased levels of carbon dioxide in the atmosphere lead to global warming. The function of carbon sequestration is to prevent excess carbon dioxide released from the burning of fossil fuels from entering the atmosphere.

Also known as *geosequestration*, this method involves injecting carbon dioxide, generally in supercritical form, directly into underground geological formations. Oil fields, gas fields, saline formations, unminable coal seams, and saline-filled basalt formations have been suggested as storage sites. Various physical (e.g., highly impermeable caprock) and geochemical trapping mechanisms would prevent the CO_2 from escaping to the surface.

Carbon dioxide is sometimes injected into declining oil fields to increase oil recovery. Approximately 30 to 50 million metric tons of CO_2 are injected annually in the United States into declining oil fields. This option is attractive because the storage costs may be partly offset by the sale of additional oil that is recovered. Disadvantages of old oil fields are their geographic distribution and their limited capacity, as well as that the subsequent burning of the additional oil so recovered will offset much or all of the reduction in CO_2 emissions.

Unminable coal seams can be used to store CO_2 because CO_2 is absorbed by the surface of coal. However, the technical feasibility depends on the permeability of the coal bed. In the process of absorption the coal releases previously absorbed methane, and the methane can be recovered (enhanced coal bed methane recovery). The sale of methane can be used to offset a portion of the cost of the CO_2 storage. However, burning the resultant methane would produce CO_2, which would negate some of the benefit of sequestering the original CO_2.

Saline formations contain highly mineralized brines and have so far been considered of no benefit to humans. Saline aquifers have been used for the storage of chemical waste in a few cases. The main advantages of saline aquifers are their large potential storage volume and their common occurrence. The major disadvantage of saline aquifers is that relatively little is known about them compared to oil fields. To keep the cost of storage acceptable, geophysical exploration may be limited, but resulting in larger uncertainty about the aquifer structure. Unlike storage in oil fields or coal beds, no side product will offset the storage cost. Leakage of CO_2 back into the atmosphere may be a problem in saline aquifer storage. However, current research shows that several *trapping mechanisms* immobilize the CO_2 underground, reducing the risk of leakage.

For well-selected, designed, and managed geological storage sites, IPCC estimates that CO_2 could be trapped for millions of years, and the sites are likely to retain over 99% of the injected CO_2 over 1000 years.

The U.S. Geological Survey (USGS) has been studying geological options for storing CO_2 in depleted oil and gas reservoirs, deep coal seams, and brine formations (source: www.energy.er.usgs.gov/health_environment/co2_sequestration/).

USGS research has focused on these themes:

- Characterization of geological and geochemical factors controlling the capacity to store CO_2 in geologic formations
- Identification of potential reservoirs for geological CO_2 sequestration
- Characterization of the geological processes that operate in natural and manmade analogs of CO_2 storage reservoirs, including high CO_2 content natural gas accumulations, oil reservoirs undergoing long-term enhanced oil recovery with carbon dioxide, and natural gas storage reservoirs.

USGS research has produced the following knowledge base:

- A geographic information system map comparing the location of major CO_2 sources with the size, location, and type of potential storage reservoirs maps.
- New measurements of the solubility of CO_2 in brines.
- New methods of assessing the CO_2 sequestration capacity of geological formations.

Potential USGS products are:

- New tools/approaches that can be used to study CO_2 sequestration in sedimentary basins
- Storage capacity assessments for major U.S. sedimentary basins

Methodology for the quantitative assessment of total geological reservoir capacity may prove applicable worldwide, allowing international assessment of carbon sequestration capacity. Details of carbon dioxide sequestration are beyond the scope of this book.

REFERENCES

[1] Rogiers JC. The use of rock mechanics in petroleum engineering: general overview. In: Hudson JA, Hoek E, editors. Comprehensive rock engineering, vol. 5. Pergamon Press Ltd., Oxford; 1993. p. 605–16.

[2] Selmer-Olsen R, Broch E. General design procedure for underground openings in Norway. In: Norwegian hard rock tunnelling, publication No. 1 of the Norwegian Soil and Rock Engineering Association. Trondheim: Tapir Publishers; 1982. p. 11–18.

[3] Froise S. Hydrocarbon storage in unlined rock caverns: Norway's use and experience. Tunnel Undergr Space Technol 1987;2(3):265–8.

[4] Lindblom U. City energy management through underground storage. Tunnel Undergr Space Technol 1990;5(3):225–32.

[5] Jansson G. Underground storage of oil and gas. Undergr Space 1983;7:275–7.

[6] Winqvist T, Mellgren KR. Going underground. Stockholm: Royal Swedish Academy of Engineering Sciences; 1988. p. 177.

[7] Glamheden R, Lindblom U. Thermal and mechanical behaviour of refrigerated caverns in hard rock. Tunnel Undergr Space Technol 2002;17:341–53.

[8] Aoki K, Hibiya K, Yoshida T. Storage of refrigerated liquefied gases in rock caverns: characteristics of rock under very low temperatures. Tunnel Undergr Space Technol 1990;5(5):319–25.

[9] Ikaheimonen P, Leinonen J, Marjosalmi J, Paavola P, Saari K, Salonen A, et al. Underground storage facilities in Finland. Tunnel Undergr Space Technol 1989;4(1):11–15.

[10] CMRI. Underground storage of crude oil, natural gas and petroleum products, an interim report of CMRI submitted to OIDB. New Delhi, India; 1993. p. 194.

[11] Moberg, S.H. Storage of heavy fuel oil in rock caverns during three decades. Proc of the first International Symposium on Storage in Excavated Rock Caverns, Rockstore' 77, 1977, Vol 1, Bergman (Ed.), Pergamon, Oxford, pp. 149–55.

[12] Calminder A, Hahn T. Recent developments in underground storage techniques. ISRM symposium, Aachen, Germany, AA Balkema, Rotterdam; 1982. p. 893–901.

[13] Kiyoyama S. The present state of underground crude oil storage technology in Japan. Tunnel Undergr Space Technol 1990;5(4):343–9.

[14] Sagefors I, Calminder A. Polytank underground liquid storage. In: Proceedings rock store, vol. I. Sweden, Pergamon Press, Oxford; 1980. p. 267–72.

[15] Calminder A, Sagefors I, System WP. Innovative underground storage for heavy crudes and sensitive petroleum products. Undergr Space 1984;8:31–35.

[16] Vered-Wiess J, Flexir A, Aisentein B, Tabary J. A new approach for LPG (propane-butane) underground storage in lined caverns above water table (Mesilat Zion, Israel). In: Proceedings of the international symposium of subsurface space. Rock Store-80, vol. 2. Pergamon Press Ltd., Oxford; 1980. p. 415–21.

[17] Dowding CH. Seismic stability of underground openings. In: Proceedings international symposium on storage in excavated rock caverns. Rockstore 77, vol. 2. Oxford, England: Pergamon Press; 1977. p. 231–8.

[18] Barton N. Seismic design concept in Q-system, training course on rock engineering. New Delhi: CSMRS; 2008. p. 404–10.

[19] Yamahara H, Hisatomi Y, Morie T. A study of the earthquake safety of rock caverns. In: Proceedings of the international symposium on storage in excavated rock caverns. Rockstore 77, vol. 2. Oxford, England: Pergamon Press; 1977. p. 377–82.

[20] Aydan O, Ohta Y, Genis M, Tokashiki N, Ohkubo K. Response and earthquake induced damages of underground structures in rock mass. J Rock Mech Tunnel Technol New Delhi 2010;16(1):19–46.

Chapter 10

Civic Facilities Underground

Come forth into the light of things, let nature be your teacher.

William Wordsworth

10.1 GENERAL

Civic facilities are part of standard city planning. To reduce congestion on the surface and to use surface land for other purposes, civic facilities such as sewage treatment plants, libraries, and shopping malls are being planned and constructed underground.

Hence, a current trend throughout the world involves the nontraditional utilization of underground space for civic construction. Apart from the various uses of underground space described in earlier chapters, this chapter is mainly devoted to underground civic facilities such as sewage treatment plants, heating systems, swimming pools, education centers or libraries, shopping malls, sports centers, and education centers. Some of these civic facilities are covered in this chapter.

10.2 SEWAGE AND WASTE WATER TREATMENT PLANT

A sewage treatment plant in Pec pod Snezkou is the first work of its kind in Czechoslovakia to be designed and constructed as an underground facility [1]. A sewage treatment plant and a drinking water purification plant already have been put into service in Harrachov, another center in Czechoslovakia's Giant Mountains.

The demand for an odorless environment requires buffer zones of more than 200 m around a surface treatment plant. In comparison, a subsurface location demands little land and offers potential advantages for the collection and control of ventilated air. Because the subsurface plant is not visible, it does not disturb the scenic landscape. The largest sewage treatment plant, Henriksdal, in Stockholm, Sweden, operates just below modern houses. Other plants also operate in residential neighborhoods, without obvious conflicts with the residents [2].

The layout for a treatment plant often consists of a number of parallel series of tanks for the purification of sewage; in each series, the contaminated sewage undergoes purification in the tanks and is cleaned. The purification process demands a rock cavern 300 m long, with a 100 m^2 cross section. In plants built earlier, tanks were built directly toward the rock, thereby limiting desirable revisions in the purification process. However, this lack of flexibility can be overcome through more farsighted design and construction. In several plants, additional caverns with complementary treatment lines have been built during normal plant operation.

The even temperature in the controlled climate is advantageous for the treatment process and also creates conditions for a favorable working environment, provided that requirements for light and ventilation are carefully satisfied. Studies to date have found the personnel in rock treatment plants to be satisfied with the working conditions. No additional salary is paid to personnel who work in underground treatment plants.

10.2.1 Case Example of Cost Comparison

While comparing the cost, two trends are worth mentioning. First, the costs of excavation both aboveground and underground have increased much more slowly than the costs of construction. Second, the cost per capita of space required for public utilities increases aboveground and decreases underground as a function of the population in an urban community. Somewhere there is a break-even point where underground installations are less expensive than aboveground installations.

The cost of land is an important factor in a cost comparison between aboveground and underground utilities. In 1978, underground installations were economical when the cost of land was around US$ $85/m^2$ [3].

A case in point is the comparison of costs for an aboveground and an underground waste water treatment plant with a capacity of 200,000 m^3/day [3]. The main parts of the plants are listed in Tables 10.1 and 10.2.

The capital investment under comparison, expressed at the 1978 price level, is given in Table 10.3. It can be seen that the aboveground plant was about

TABLE 10.1 Schematic Description of an Aboveground Waste Water Treatment Plant (200,000 m^3/day) [3]

All Installations Aboveground
Administration building
Control and machinery building
Pumping and screening
Grit removal tank
Aeration tanks
Sedimentation tanks
Sludge treatment plant
Required area for plant = 100,000 m^2
Required area for buffer zone = 150,000 m^2

TABLE 10.2 Schematic Description of an Underground Waste Water Treatment Plant (200,000 m³/day) [3]

Installations Aboveground
Administration building
Control building
Sludge treatment plant
Installations Underground
Machinery
Pumping and screening
Grit removal tank
Aeration tanks
Sedimentation tanks
Required area aboveground = 20,000 m²

TABLE 10.3 Capital Investment for Aboveground and Underground Waste Water Treatment Plants Described in Tables 10.1 and 10.2 in Millions of 1978 US$ [3]

Investment	Aboveground	Underground
Construction costs		
Civil works aboveground	55.0	41.2
Civil works underground	—	23.0
Equipment for treatment	13.8	13.8
Ventilation	—	3.4
Total construction costs	68.8	81.4
Other costs (land excluded)	15.8	22.7
Interest during construction	11.7	13.8
Total investment (land excluded)	98.8	118.4
Cost of land	250,000 × L[a]	20,000 × L[a]

[a] The break-even point was around US$ 85/m² in 1978.

$20 million less expensive but required 230,000 m^2 more surface land. Consequently, a savings in investment was possible in communities where the cost of land was at least US$ 85/m^2. Operation and maintenance costs are about the same for aboveground plants and underground plants. A sensitivity analysis has shown that the construction cost of an underground plant can increase by 38% if the cost of rock excavation and necessary reinforcement double due to poor rock conditions [3].

10.3 SPORTS CENTER

As on the surface, a sports center is a conglomeration of various sports facilities. A sports center underground is also planned for more than one sport. A project in Italy involves construction of a sports center that can also serve as a fallout shelter. The main feature of this project is the use of three parallel caves. The first cave would contain an Olympic-size swimming pool (38 × 18 m long); the second would serve as a hockey and skating track (105 × 20 m high); and the third, the largest, would house four tennis courts and a pool. The total excavated volume of the project would amount to 105,000 m^3 [4].

Due to a lack of suitable space on the surface for a sports hall and swimming pool—the surface land available being needed for housing—it is found convenient on many occasions to use the underground space for such activities. It has been found that the underground option for such facilities has other advantages in the saving of energy and in providing a safe shelter for people in wartime.

10.3.1 Underground Ice Rink

The world's largest rock cavern for public use was developed in Gjovik, Norway, as an ice hockey rink for the 1994 Olympic winter games. Featuring a free span of 60 m, the cavern houses an ice rink that complement the city's other cavern installations, which include a public swimming pool and a major telecommunications center.

In addition to the main 60 × 90 × 25-m ice rink chamber, the Gjovik Hall includes all necessary technical facilities, changing rooms and cloakrooms, offices, and a restaurant. The entrance tunnel to the hall begins from the center of the town and provides space for turnstiles, ticket offices, a vestibule, and a "rock garden" cavern.

Placing the hall underground offers numerous advantages, such as the environmental benefits associated with avoiding crowding the surface with a huge facility and the ability to integrate the new hall with the existing underground swimming hall. In addition, underground placement offered potential energy savings and security advantages. Also, as is the case with many other subsurface facilities in Scandinavia, the new hall can serve a dual purpose as an air raid/civil defense shelter.

Construction of the hall involved drill-and-blast methods. The rock is Pre-cambrian gneiss. All exposed rock surfaces were shotcreted, and rock bolts and anchors were used as support.

10.3.2 Swimming Center

Underground swimming pool installations generally require space for a swimming pool, locker room, and shower with an admission tunnel. In case other facilities also need to be planned, the size of the cavern can be decided accordingly or a number of small caverns can be designed depending on the rock type.

Underground swimming pools have been found to be energy efficient. One such study carried out by [5] in Norway shows that a swimming pool in rock requires less energy compared to conventional swimming pools on the surface (Figs. 10.1 and 10.2). In fact, a swimming pool in rock has an annual energy

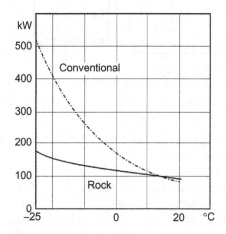

FIGURE 10.1 Energy demand as function of outside temperature. Pool temperature 28°C [5].

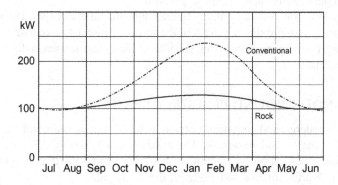

FIGURE 10.2 Energy demand as function of monthly mean temperature. Pool temperature 28°C [5].

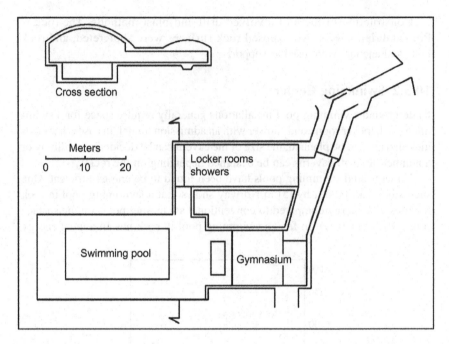

FIGURE 10.3   Layout of sports facility, including swimming pool in rock at Gjovik, Norway.

consumption for heating and ventilation that is 65% of a conventional swimming pool (28°C pool temperature).

The swimming pool in rock at Gjovik, Norway, consists of two caverns with a common admission tunnel and connection tunnels (Fig. 10.3). The largest cavern contains the swimming pools and the gymnasium. The smaller cavern contains locker rooms and showers for the swimming pools. Technical rooms are situated partly in the basement of the swimming pool, partly on the roof above the gymnasium, and in the locker rooms. The rock is limestone with a thermal conductivity factor equal to 2.5 w/m^2.

The ventilation system is similar to the conventional alternative. The capacity, however, is approximately 65% of the former. The relative humidity in the swimming pool cavern is constant at 65%. At the outer entrance a dry zone is arranged [5].

10.4 UNDERGROUND PEDESTRIAN PATH

Toronto's underground pedestrian network is a labyrinth of tunnels, basements, escalators, malls, and fountains. The system, which serves a Toronto area metropolitan population of 3.5 million, has grown to be the largest system of its kind in North America and, perhaps, the world [6].

Toronto's long, cold winters are certainly one important factor encouraging new developments to connect into the underground network. Another factor is the unprecedented growth of the city's financial district in the 1960s and 1970s. This economic growth provided the opportunity to separate pedestrian circulation from automobile circulation, with obvious advantages to each.

The present subsurface system supports 30 office towers, city hall, Union Station, two department stores, 20 parking garages, three hotels, and the stock exchange. The subsurface space contains about 1000 retail shops, numerous restaurants, two cinemas and a cineplex, and nine fountains with gardens. Street level entrances number over a hundred.

10.4.1 Operation of the System

Two Networks

The downtown Toronto system comprises two separate but connected networks. The northern network operates in a typical shopping center fashion, with a department store at each end and a retail mall in between. The northern anchor is the Toronto Eaton Centre Galleria, Canada's most successful shopping complex and the city's main tourist attraction (50,000 tourists visit the complex daily). The southern anchor is Simpson's Department Store. The mall is open to the public during the hours of subway operation, that is, 6:00 a.m. to 2:00 a.m. The Galleria stores are open late, while the department stores have more limited hours. The bulk of the center's retail business is conducted on Saturday.

The southern network interconnects the major bank towers of the financial district. Shops and stores serve the weekday office population essentially between the hours of 10:00 a.m. and 2:00 p.m. It now has become important for new projects in the financial district to offer their prospective tenants access to the amenities of the underground network.

Security

The network is privately owned and is subject to the control of private police. Although the mall and arcade are open to the public and seem to be public, in reality they are private property. Spaces and tunnels in the system are treated as a private utility. The responsibility for security rests with the property owners. The private security system makes extensive use of video monitoring and radio communication.

Orientation

Because the system is complex and can be confusing to the occasional visitor, efforts are underway to improve directional graphics. Daily users get to know the system and seem to find their way around quite easily.

The underground network in Toronto has secured economic and environmental benefits, including [6]:

- Increases in property values and land utilization
- Separation of pedestrian and automobile traffic
- Linking of transportation modes
- Reduction of surface congestion
- Protection of pedestrians from inclement weather
- Coordination of the infrastructure
- Improved atmosphere for pedestrians

The underground pedestrian and utilities network in the campus of Laval University, Quebec, Canada, covers well over 4 km of tunnels used by students, teachers, and personnel. Small electric vehicles are used to carry mail and merchandise through the network, which links 20 separate buildings. Frequent visits to this underground network revealed that it is very popular in the winter and on rainy days. The decor is strictly functional, and locally pipes are visible side by side or overhanging the pedestrian, as shown schematically in Figure 10.4. The walls of

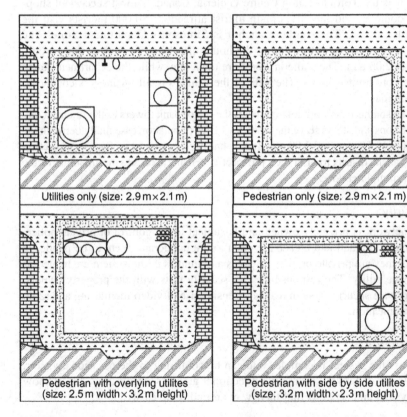

FIGURE 10.4 Different sections of pedestrian and utility tunnels on the Laval University campus in Quebec City [7].

the tunnels are covered with graffiti of all sorts. The university administration considers the graffiti as rather an advantage because it is believed that the relative absence of graffiti on surface buildings and in corridors of the upper floors is related to the extensive graffiti underground [7].

Next to this network is the utilities network, which shares some hundred meters with the pedestrian tunnels. Water, heating, and electricity serve all of the buildings on the campus, and additional cross-sectional space is available for future expansion.

Each urban area has its own unique circumstances that will influence the potential application of the Toronto, Montreal, and Quebec ideas of pedestrian network. These include urban density, local geology, building configuration, development policies, and environmental standards. A thorough analysis of local needs and capabilities is essential to sound planning for subsurface use.

10.5 SHOPPING MALL

The first underground shopping mall in Japan was completed in 1932. By 1985, a total of 76 underground malls had been built. The total area of all of these underground shopping malls is 818,000 m². Of this area, approximately 27.4% consists of public walkways, etc.; 29.5% consists of shops; 26.6% consists of vehicle parking spaces; and 16.4% is taken up by machines, etc. These figures indicate that areas for access and parking equal to the area of shops have been ensured [8].

Because underground malls are usually built around stations where large numbers of people gather, they extend under both the station squares and the surrounding roads. The following conditions are imposed on the development of underground malls [8]:

- Development of roads around the station and the station square must have been completed and there must be no plans for further development.
- Development of the underground mall must be necessary for pedestrian safety due to overcrowding of aboveground roads.
- An underground car park should be provided to give cars access to the station and underground mall without disturbing aboveground road transport.
- An underground sidewalk and car park should be constructed as part of an urban development plan, and the provision of an underground mall with shops should be regarded as a supplementary amenity to the aforementioned public facilities.
- The organization in charge of the construction and management of the underground mall should be a quasi-public organization with at least one-third of the capital paid by a public organization, such as a municipal authority.
- The Building Standards Act and the Fire Services Act should be strictly applied to facilities under roads and station squares. Some important measures to be applied in this regard in developing underground malls are as follows.
  - Underground pavements must be at least 6 m wide.
  - Noncombustible materials must be used for all interior work.

- Shops must be built in such a way that they can be divided into areas of less than 200 m² and surrounded by fire prevention zones.
- Restaurants and other establishments using gas must be located in separate areas from other shops and must be surrounded by fire prevention zones.
- When an underground mall is connected to the basement floors of other structures, these basement floors must be equipped with fire prevention measures similar to those of the underground mall. Stairs to ground level must be provided at connection points with fire prevention zones.
- When an underground station is provided with an underground mall, it must be equipped with at least two stairways leading directly to ground level from the platform level; these stairs must be completely protected by fire prevention zones.
- Sprinklers, automatic fire warning equipment, a broadcasting system, guiding lights, fireplugs with connecting hoses, smoke discharge units, and an emergency power unit (either battery or independent power generating unit) must be provided and must be centrally controlled by the disaster prevention center.

Various components required for the disaster prevention for the Yokohama station in Japan are listed in Table 10.4. Table 10.4 can serve as a guide for the disaster prevention plan in shopping malls.

TABLE 10.4 Components of Disaster Prevention Plan for Yokohama Station Underground Shopping Mall in Japan [8]

Electrical facility	Emergency power unit	
	6.6 kV	1000 kV (1 unit)
	Battery	100 V
		500 AH, 600 AH (1 unit)
Water supply facility	Water tank, 620 ton capacity (1 unit)	
Fume exhaust facility	Natural fume exhaust—eight vents	
Communication and signaling facility	Loudspeakers and emergency telephone system covering the entire facility, plus a supplementary facility for radio communication	
Disaster prevention facility	<ul><li>Fire alarm with smoke and heat sensors (1800 locations)</li><li>Indication on large instrument panels at the central control room of the disaster prevention center (CCR-DPC)</li><li>Monitoring with an industrial TV system</li><li>Automatic reporting to police and fire departments by linked coaxial cables</li></ul>	

TABLE 10.4 Components of Disaster Prevention Plan for Yokohama Station Underground Shopping Mall in Japan [8]—cont'd

- Reporting to DPC by emergency telephones (19 sets)
- Emergency lights (500 locations, 10 luxes each), guiding lights (80 units)
- Outdoor hydrants (38 locations)
- Sprinklers (3234 locations)
- Bubble-type fire-fighting facility (mainly at car parks)
- Inert gas fire-fighting facility (electrical room and others)
- Fire-fighting water storage (aboveground)—130 m^3
- Fire-fighting water storage (B$_2$ floor)—92 m^3
- Fire-fighting entrance—two locations (aboveground on the national highway)
- Connecting water pipe—three locations (aboveground)
- Emergency gas shut-off valve (linked to CCR-DPC) with a double-check system
- Gas sensors at 46 locations (tenant stores) and alarm units
- Opened gas valve indicator for each tenant store and safety check valve (linked to CCR-DPC)
- 12 gas leakage sensors on gas lines for each fire prevention zone (linked to CCR-DPC)
- Fire bulkhead closing facility (remote controlled by CCR-DPC)
- Natural fume-exhaust opening facility (remote controlled by CCR-DPC)

10.6 UNDERGROUND RECREATIONAL FACILITIES

In most cases, because facilities for public entertainment and cultural activities cannot spread upward, the only alternative is to use underground space. Utilization of underground space for public entertainment and cultural activities has advantages such as little or no occupation of surface space, energy savings, environmental protection, and economic benefits. There are a number of other reasons as follows for building these facilities underground [9].

- Because people generally stay in entertainment facilities for only short periods of time, there is no such a great need to provide direct views of the outside. Because people using these facilities often have enough time to enjoy outdoor activities in the sunshine and natural air, for example, after or before a concert or movie, the surface surroundings should be as attractive as possible, combining attractive landscaping with well-planned roads. In addition, with proper ventilation, moisture protection, and dehumidification inside the space, subsurface entertainment facilities can have the same effect as similar facilities on the ground. Moreover, a well-designed interior can render an underground facility very attractive.

- Underground spaces can be constructed at those points where streams of people from administrative offices, schools, and factories are concentrated. This arrangement is convenient for workers, shoppers, and students. In addition, such an arrangement can maximize economic benefits. Experience in Shanghai and Hangzhou in China has shown that these facilities have a high rate of utilization and that investment costs can be recovered in 1 to 2 years [9].
- Low construction costs. Underground facilities for entertainment and cultural activities create no interference on the surface. Because there is no need for land requisition or building demolition, such facilities cost less and can be constructed more quickly than similar surface facilities. There is flexibility in selecting building sites: the facilities can be built separately or as an extended basement. The span and height are determined in accordance with the topographical and geological conditions in order to make full use of the mechanical properties of the rock and to reduce construction costs. If excavated soil and waste from underground construction are used for building material production, such as aggregate or raw material for brick, even more benefits can be gained.
- Underground cultural and entertainment facilities can be adapted for use during peace or war. Because such facilities are well located and have few internal installations, it is relatively easy to convert them into shelters during a war. In fact, many of China's underground facilities for entertainment and cultural activities formerly served as such shelters. With unified planning and guidelines for use during peace or war, the construction, management, and utilization of underground facilities can all be coordinated. Such a system provides economic benefits and offers proper maintenance of the shelters in peacetime. Thus, good utilization makes for good maintenance.

Typical examples of underground sightseeing attractions, recreational facilities, amusement halls, and so on developed in China have been nicely discussed by [9]. It is important to mention that these facilities are well received by people.

REFERENCES

[1] Uher P. Czechoslovakia's first underground sewage treatment plant. Tunnel Undergr Space Technol 1987;2(3):275–8.

[2] Jansson B. Water and sewage in caverns and tunnels: thirty years' experience in Sweden. Tunnel Undergr Space Technol 1989;4(1):17–51.

[3] Isgaard E, Harlaut A. Underground public utility systems: why and where? Undergr Space 1983;7:306–7.

[4] Pelizza S. Civil applications of underground works in Italy. Tunnel Undergr Space Technol 1987;2(3):283–5.

[5] Dorum M. Energy economy in sports halls and swimming pools in rock. Norwegian hard rock tunnelling, publication No. 1, Norwegian Soil and Rock Engineering Association. University of Trondheim, Norway: Tapir Publishers; 1982. p. 19–23.

[6] Barker MB. Toronto's underground pedestrian system. Tunnel Undergr Space Technol 1986;1(2):145–51.

[7] Boivin DJ. Underground space use and planning in the Quebec City area. Tunnel Undergr Space Technol 1990;5(1/2):69–83.

[8] Tatsukami T. Case study of an underground shopping mall in Japan: the east side of Yokohama Station. Tunnel Undergr Space Technol 1986;1(1):19–28.

[9] Xu SS. Development and use of underground space for entertainment and cultural activities in the People's Republic of China. Tunnel Undergr Space Technol 1987;2(3):269–73.

Chapter 11

Underground Structures for Hydroelectric Projects

Out of site is out of mind.

Cited by Professor Charles Fairhurst

11.1 INTRODUCTION

Minimization of entropy is a new principle of planning and development. Thus global entropy planning is a must in the 21st century to stop global side effects such as global warming and climate disturbances [1]. It means simply that the use of heat energy and uncertainties must be minimized with an increase in the efficiency of technologies for global good health, like in all living species. Thus energy saved due to an increase in efficiency is energy generated. The entropy analysis will reveal the truth (entropy rise ≈ total heat energy created in life span of a project divided by the absolute temperature). It may put a check on all lies. We should pray to rise above cheating. Most heat energy-based power-generating plants (atomic + coal + gas + oil etc.) cause a very high entropy rise and corresponding side effects, whereas hydroelectric projects generate almost no heat or entropy rise. A society cannot survive without good intentions. Unethical energy generation cannot be sustainable at all for a long period globally and cannot generate more self-reliance. The energy cannot give lasting happiness, for lasting happiness is evolving deep within all healthy species. The top priority of all nations should be to increase their electrical power potential for rapid healthy economic development to make our life comfortable. Electricity is a key requirement of modern civilization. Thus the interest of city planners is increasing in underground space technology. The world is going for automation in tunneling technology in economical ways. It is a humble presentation. Wise readers should kindly draw their own conclusions. Thus our anxiety of entropy may be reduced.

The efficiency of hydroelectric generation is as high as 80%. The maximum efficiency of a thermal power plant is 50% and may be as low as 20% in most plants. Solar energy cells have the lowest efficiency of 15%. The efficiency of nuclear power plants is also about 15% [2]. Further, commitments to reduce greenhouse gas emissions under the Kyoto protocol provide new incentives for developing hydroelectric dams. There should be productive harmony between us and nature.

Only well-protected, ethical energy generation is sustainable globally for a very long time as the same will be acceptable to people happily all over the world. Hydroelectricity is the least expensive, cleanest, safe, hygienic, and healthy energy from an ever-renewable, natural, abundant resource: water. Although the initial investment is high and construction time is too long, there are no unhealthy side effects such as from thermal or atomic power projects. For safety, atomic power plant projects need to be designed to thwart any potential attacks launched upon it. As a result, the perception is that hydroelectric generation is going to be used for a long time. Atomic power plants should be located far away from the tsunamis, say 50 m above mean sea level. As such, public opinion needs to be mobilized in cities. It is catching up rapidly. Energy generation is a highly profitable business now.

High dam storage projects protect us from floods and are useful for irrigating crops through a network of canals and supplying clean drinking water to thirsty nearby mega cities. Mega interriver linking projects are going to be the boon of the future by creating a vast network of very long canal tunnels, canals, dams, hydroelectric projects, drinking water networks, pipe lines, and so on in some countries. The strengthening of dams is an ongoing process for their safety. River basin water, mother of all life, has to be regulated through the creation of safe dams to draw maximum life support from this precious resource of water.

High dams, that span across big rivers, can harness the super destructive energy of recurring dangerous floods to irrigate thirsty fields and generate hydroelectricity to serve the suffering nations. The free trade of sediments between adjoining hilly nations is raising river bed levels in flood plain areas. As a result, embankments of thousand-kilometer lengths but of smaller height (with unmaintained sluice gates) are breached easily. As such, the flood plain area increases slowly due to ever-rising river bed levels. High dams with reforestation of vast catchment areas can stop all this yearly loss of life and property and instability of governments. China has built more than 100,000 small and large dams. A very large number of small dams are needed to replace a high dam to produce the same electricity or irrigate the same area of fields. Where are the best sites for so many small dams in fragile mountains?

Asia is the fastest-growing economic region in the world. Table 11.1 lists plans for generating hydroelectricity in 14 Asian nations [3]. The high economic growth of China and India demands 200,000 MW electricity in each country by 2020. A three gorges dam producing 17,000 MW is a historical achievement. India and Japan are building dams with pumped storage on the downstream in which stored water is pumped back into an upstream reservoir at night in order to generate electricity in the daytime when it is most needed. Many plants are developed by private companies on build–operate–transfer arrangements. Malaysia is developing mini hydro projects on a build–own–operate concept. Interaction among private, academic, and research institutions has increased after privatization. Norway is generating 99.8% of its electricity from hydropower

TABLE 11.1 Current Hydropower Contributions and Planned Additions in Selected Asian Countries [3]^a

Country	Installed Hydro Capacity (MW)	Percent of Total Capacity	Planned Hydro Additions through 2010 (MW)
Bangladesh	230	10	0
Hong Kong	0	0	1200
India	22,000	21	16,000
Indonesia	2100	24	4000
Japan	38,500	22	1000
Malaysia	1500	22	1500
Myanmar	250	32	600
Philippines	2200	28	1700
People's Republic of China	40,800	18	95,000
South Korea	2500	9	4000
Sri Lanka	1100	83	55
Taiwan	4100	24	1.100
Thailand	2300	19	2500
Vietnam	1700	63	6100
Total	119,280		134,755

^aData courtesy of the international consulting firm Arthur D. Little, based in Cambridge, Massachusetts.

projects on safe selected rivers only. Tunnels of 4000 km length have been excavated for this purpose, and Norway has 200 underground powerhouses in hard metamorphic rocks, which is 40% of the world's 500 underground powerhouses. The pressure tunnels and shafts are unlined even under water heads of 1000 m. The newly developed air cushion surge chamber has replaced conventional vented surge chambers. None of the unlined pressure tunnels and shafts with water heads varying from 150 to 1000 m, which have been constructed in Norway since 1970, has shown unacceptable leakage [4]. Norwegian engineers deserve all congratulations.

Hydropower is capital intensive. Its initial cost is the highest of all other projects, but operation cost is the least of all. The cost of input is zero. Hydropower will be most economical after, say, 20 years, as the inflation rate is merely 1 to 2%, whereas the inflation rate for thermal and uranium/thorium is more than 15% per year. Further, it is costly to build special plants for the removal of sulfur dioxide from gases out of thermal power plants, which is essential in protecting fragile ecological systems. The cost of hydropower is slightly more than the cost of thermal power and is about half the cost of nuclear power just after commissioning of projects. Ultimately hydropower will be most economical as water is free of cost. Further, gas hydrates are likely to be the new source of electrical energy all over the world for the next thousand years. Nature has been benevolent in distributing gas hydrates along the shallow sea coasts all over the globe (www.geology.usgs .gov). We should ensure that inputs are locally available in abundance whatever the strategy of energy generation for economic development may be.

Engineers are planning a series of run-of-the-river hydroelectric projects where river beds are steep, such as in the upper Himalayan countries and China, as they are people-friendly and eco-friendly areas. This trend is picking up gradually as the cost of underground construction starts comparing with the inflating cost of surface construction. The successful completion of engineering marvels, vast hydroelectric projects, is the proof of the confidence, willpower, and wisdom of modern engineers and geologists.

It may be startling to engineers that head race tunnels (HRT) with PCC lining have been functioning since 1980 nonstop in the Himalayas with no problem of rock faults and underground instability. Planners are concerned with minimizing geological hazards in their vast plans of hydroelectric projects. Figures 4.3 and 4.4 in Chapter 4 may be helpful.

With advances in tunneling technology, hydroelectric power projects with nearly all underground structures are more economical, safer, and durable than surface structures, particularly in a rugged terrain, seismic hilly areas prone to landslides or snow avalanches, and strategic areas subjected to wars. The cost of stabilizing slopes such as dam abutments, spillway slopes, and steep cuts behind power stations is very high.

In developing countries there are not many underground powerhouses. However, a large number of them are likely to be constructed in the near future as part of multipurpose river valley projects. Jaeger [5] describes rock engineering problems of underground powerhouse complexes with the help of many case histories.

In modern times, the first underground powerhouse was constructed at Snoqualmie Falls near Seattle in the United States in 1898–1899. It may be interesting to note that the construction of underground powerhouses went on with an accelerated pace only after World War II.

The ASCE Symposium on Underground Chambers held in 1971 contains useful articles. A case history on the Churchill powerhouse by Benson and colleagues [6] is comprehensive and worth reading. Figure 11.1 shows the current trend in planning hydroelectric projects. Without raising the height of a dam, the head of water is increased by going underground.

Despite a great deal of experience generated over the years, the design and construction of underground structures are largely based on precedence and experience. The construction of new tunnels and caverns is thus undertaken

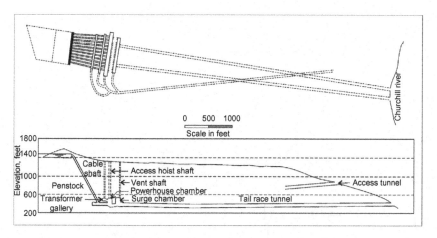

FIGURE 11.1   Layout of Churchill Falls hydroelectric project [6].

by extending previous experience and using observations and the skill of construction engineers.

Conservatism thus appears to be the guiding philosophy. Modern underground planning and construction are based on the design-as-you-go philosophy, which has proved to be very effective and economical [7].

11.2 RECENT DEVELOPMENTS IN PLANNING OF HYDROELECTRIC PROJECTS

With confidence and experience gained in tunnel engineering, there have been major innovations in the planning of hydroelectric projects. As mentioned earlier, with the same height of a dam, the head of water can be increased by taking penstocks deep in the ground. In this way, the head of water up to 500 m or even 1000 m becomes possible for hydroelectric generation. Figures 11.2 and 11.3 show some typical schemes of underground development of powerhouse complexes [5]. Planning a surge chamber upstream of a powerhouse is generally considered best. If the hill slope is very stable, a power station may be built on the ground itself.

Another recent development in powerhouse design is use of a compressed air surge chamber (Fig. 11.4). It is located deep inside massive rock, which is of low permeability, but the air hydraulic permeability of the rock mass should be as low as 10^{-7} cm/s [8]. Two such air surge chambers have been built in Norway and are doing well.

Dolcetta [9] reported a novel design of very high underground power station in a poor quality of rock mass. The stability of the underground powerhouse was a problem due to the plastic zones formed around the opening. The problem was solved by designing a thick R.C.C. slab acting as a brace between walls. Long

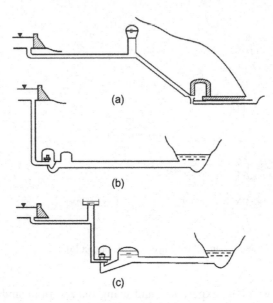

FIGURE 11.2   Types of underground power stations: (a) head race arrangement, (b) tail race arrangement, and (c) intermediate solution [5].

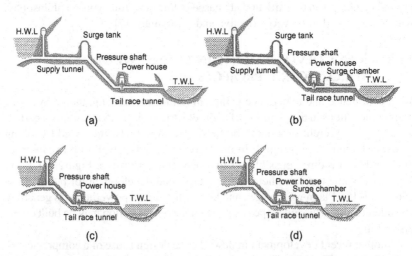

FIGURE 11.3   Typical schemes of underground development: (a) upstream surge tank, free flow tail race; (b) two surge tanks system; (c) no surge tank; and (d) downstream surge tank [5].

cable anchors were also considered essential to control the squeezing of the rock mass. The paper describes a special sequence of excavation that ended with an installation of supports and construction of a R.C.C. brace without any excavation hazard. All these figures show that the design and construction of underground powerhouses are challenging tasks.

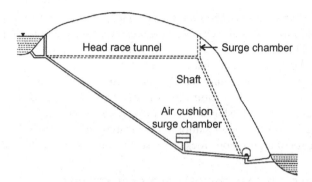

FIGURE 11.4   Utilization of compressed air cushion surge chamber as compared with conventional shaft design [8].

Another development that is picking up nowadays is building a pair of (underground) power stations. One is as usual for power generation during peak periods and the other is for pumping water back into the reservoir during off periods. The idea is to use hydroelectric power systems to meet peaking demand and use thermal stations for normal demand.

Mistakes have been committed in the past in the planning of underground systems of hydroelectric projects.

- The alignments of long power tunnels have not been fixed after proper and purposeful geological exploration. Surface geological features and surface rock mass classification have proved misleading. Consequently, many disastrous problems were faced in tunneling.
- In squeezing grounds, selection of the size of a power tunnel is important. Initially, one big tunnel was excavated, as it became difficult to drive, so three tunnels of smaller diameter were then driven [10]. This was because there were fewer supports, the bridge action period was greater, and heaving of the floor was limited.
- The underground powerhouse was located at one side of the river and not high enough over the water level in the river. Consequently, during flood, water entered into the cavern through the tail race tunnel (TRT) while its excavation was going on. Rock masses in the roof and walls were submerged for a couple of weeks before water could be pumped out. Fortunately, the cavern remained stable.
- Seepage through the dam foundation was increased after a major earthquake, as permeability of the jointed rock mass increased drastically during shearing. However, the permeability of microfissured rock was reduced due to the closure of joints beneath the foundation due to thrust on reservoir filling, thereby making a grout curtain redundant, which led to a dam failure.
- A study (on the basis of 11 years of monitoring the Chhibro cavern of the Yamuna Hydroelectric Project) has shown that ultimate roof support

pressure for water-charged rock masses with erodible joint filling may rise up to six times the short-term support pressure. Damage to the support system during an earthquake of 6.3 magnitude is not appreciable except near faults with plastic gouge material [11].

- Very high support pressures may be generated by a reduction in the modulus of deformation due to saturation of rock mass around HRT, TRT, and penstocks, etc. [12]. Mehrotra [13] observed that modulus of deformation is actually very low after saturation compared with that in a dry condition in the case of argillaceous rocks (claystone, siltstone, shale, phyllite, etc.).

11.3 TYPES OF UNDERGROUND STRUCTURES

There are basically the following underground structures within a mountain for a hydroelectric project, which is a modern trend in the 21st century.

- Desilting chambers to extract coarser particles of sediment of water (say > 0.15 or 0.2 mm in size), as otherwise turbine blades will be damaged. Tunnel intakes are also popular. Swamee [14] has suggested design criteria for desilting chambers.
- Head race tunnel (or a canal tunnel in case of a small hydroelectric project) to carry water to the machine hall. Diversion tunnels divert river flow during construction of a dam to dry the site.
- Surge shaft(s) to absorb hydraulic energy due to water hammer on sudden closure of turbines.
- Penstocks to carry water to each turbine individually under a large hydraulic head of water.
- Machine hall cavern and transformer cavern to house turbines and generators for generating hydroelectricity.
- Tail race tunnel to discharge water from the machine hall to downstream of the river with a steeper slope.
- Vertical shaft spillways in reservoir upstream of the river to discharge excess flood water during a heavy rainy season automatically in the cases of earth dams.
- Tunnels for pumping water from the pump storage reservoir back to the upstream reservoir during a nonpeak period.

11.4 PRINCIPLES OF PLANNING

As we can see, underground structures play an important role in modern hydroelectric projects. The principles of planning above underground structures are as follows [15,16; also see Section 4.4].

- Electrical power is generated in proportion to the product of the river water discharge and the hydraulic head between reservoir water level and the turbines.

- The size of the machine hall depends on the generation of electrical power and number of turbines. The machine hall should be deeper in the hill for protection from landslides. The powerhouse may be located on a stable slope in the case of hard rocks, if it is not prone to landslides and snow avalanches. The pillar width between adjacent caverns in competent non-squeezing ground should be at least equal to the height of any cavern or the average of widths (B) of caverns, whichever is larger. Displacement of the walls of the machine hall should not be high at haunches (say >6 or 7 cm) so that a gantry crane can travel freely for maintenance of machines. Swelling ground should also be avoided, which may contain montmorillonite, etc. In case a poor rock mass is expected in the lower parts of a cavern, caverns should be shifted elsewhere to a better rock mass (GSI > 35). The axis of the cavern may be oriented to be nearly perpendicular to the (active) fault zone and strike of the critically oriented joint set, if it is economical. A cage of drainage galleries may be planned to keep the rock mass dry all around the roof of caverns (machine hall and transformer hall), which are located in pervious/soluble and water-charged strata (see Fig. 11.5). A minimum rock cover above the crown of caverns should be 3B preferably, where B is the width of the cavern. A vertical surge shaft is planned sometimes up to the top of the hill surface. In case of weak rock/soil slopes around the shaft, the cut slope angle should be selected to be stable according to the value of Q_{slope} [17] or stabilized with rock anchors or cable anchors with drainage holes. Smooth blasting should be used to reduce the blasting vibrations, which do not cause damage to nearby buildings or village huts.
- The size of the desilting chamber or settling basin (rock cover H > B) depends on the size of particles needed to be extracted from the sediment from intake. Some rivers, such as the Ganges in the Himalayas, have a very high percentage of feldspar and quartz particles (of hardness of 7), which

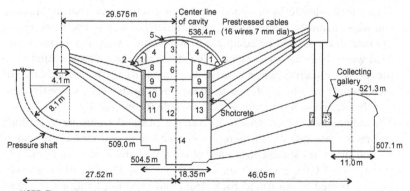

NOTE: The numbers in the zones in the cavity indicate sequence of operation/excavation

FIGURE 11.5 Benching and supporting rock in a powerhouse cavern.

damage the blades of the turbines (of weldable stainless steel) quickly. In such cases, desilting basins may be longer. The (pillar) spacing of desilting chambers is of critical importance. Turbines are closed when the silt content is more than the design limit in the river during heavy rains.

- The rock cover should be more than three times the size of openings (power tunnels). The overburden pressure should counterbalance the water head in the head race tunnel and penstocks, etc. However, the rock cover should be less than $350\,Q^{1/3}$ m for $J_r/J_a < 0.5$ to avoid squeezing conditions. Further, the overburden should also be less than 1000 m and $N > 1.5$ for $J_r/J_a > 0.5$ to avoid rock burst conditions.

- Designers have to select the velocity of the flow of water (say 4–6 m/s) in a head race tunnel. A higher velocity means more head loss and more abrasion of concrete lining. Less velocity, however, leads to a bigger diameter of the head race tunnel, which increases cost and the time of construction. Thus an economical diameter of the head race tunnel is found out by a trial-and-error approach. However, the overexcavated diameter of the tunnel should be equal to the actual diameter plus expected tunnel closure, which can be quite high in squeezing and swelling grounds. There is no need for hoop reinforcement, despite high hoop tensile stresses in lining within strong rocks, as cracks (<3 mm) are self-healing. The center-to-center spacing of HRTs and TRTs (of width B) should be 2B in (competent) nonsqueezing grounds, and preferably 6B in squeezing ground conditions for stability of the rock pillars. The width of the tunnels should preferably be limited to 6 m in severe squeezing conditions, as support pressure increases with width (for $Q < 4$). Unlined HRTs and TRTs may be planned in unerodible and hard rock masses ($B < 2\,Q^{0.4}$ m and $H < 275\,N^{0.33}\,B^{-0.1}$ m, $N = Q$ with SRF = 1, B is the tunnel size in meters), where it is more economical than tunnels lined with PCC. Segmental concrete lining, which is erected by TBM, may be used as the head race tunnel after primary and secondary grouting. PCC lining and the steel liner in penstocks should be designed considering actual rock sharing pressure and water hammer, along with maximum head of water. The thickness of the steel liner should be adequate to withstand external seepage water head when the penstock is empty. The steel liner should have stiffeners to prevent its buckling when empty. Alignment of the head race and tail race tunnels may be nonlinear to reduce overburden (H), squeezing and rock burst, and to have more adits if feasible (economically).

- The alignment of underground structures is planned according to geology and rock mass quality. If there is one band of sound strata, the underground structures should be aligned along this sound strata (with generally $Q > 1$, $E_d > 2$ GPa, except in shear zones; but $H < 350\,Q^{1/3}$ m). Penstocks may be turned along sound rock strata and need not be along the head race tunnel. The general alignment should preferably be perpendicular to shear zones/faults/thrusts zones. The thickness of the intrathrust zone should be minimized, if any. The caverns should not be parallel to strike of continuous joints but be parallel to major horizontal in situ stress if possible.

- In fact, designers cannot plan accurately in the beginning as a few drifts or drill holes give no idea of complex geology in tectonically active regions. We know complete geology about a project only when the project is over in a complex geological setup.
- A good reservoir site is located where the width of the reservoir expands upstream and preferably downstream of a confluence of rivers so that the volume of river water storage is maximum. Steep gorges such as in the Himalayas need high dams for reasonable reservoir storage (~10,000 MCM). A high dam stores water in the rainy season. Engineers are planning run-of-the-river schemes along rivers with steep bed shapes but they do not store more water for power generation but are good ecologically (China, Himalayas). The life span and the capacity of a reservoir increase with the height of a dam and so play an important role in planning river valley projects [18]. The delta formation at the confluence of a river and dam reservoir increases its life span by arresting the silt load.
- The whole plan should be feasible geologically and economically. Several layouts are made and costs compared. In the 21st century, the time of completion of a project is very important, as time means money (monthly profit). The cost of construction time should be added to the total cost of the project for optimization of the overall cost. As a result, engineering geology and rock mechanics studies are very important in reducing the time of construction of underground structures, especially very long tunnels (>5 km) under a high overburden in weak rocks. In three gorge dam projects (17,000 MW) in China, much money and construction time was saved because of properly planned rock/geotechnical investigations.
- Topography or slope of the hill river bed matters. A longer head race tunnel means more hydraulic head, which gives more power.
- Underground structures are preferred in a highly seismic area or landslide-prone area or in strategic areas as they are more stable, durable, and invisible. Surface structures are prone to damage by landslides (such as in the Himalayas and Alps).
- Major tectonic features such as tectonically active thrust and fault zones (e.g., MCT, MBT, and so on in the Himalayas) should be avoided, as articulated or flexible concrete lining may be provided to deform without failure during slip along these active thrust zones in the entire life span of the project. It is interesting to know that the observed peak ground acceleration increases with an earthquake magnitude up to 7 M (on Richter's scale) and saturates at about 0.70 g up to 8 M [19].

- A major shear zone should be treated by the dental concrete up to an adequate depth beneath a concrete dam foundation. The abutment should be stabilized by grouted rock bolts or cable anchors if the wedge is likely to slide down. In fact, a roller compacted concrete dam may be more economical than a conventional gravity concrete dam on weak rocks. Doubly arched dams will be economical in hard rocks and stable strong abutments. Stepped spillways on stable rock slopes are preferred nowadays for reducing the hydraulic energy of rapidly flowing water. The double rows of grout holes should be used to provide a tight grout curtain in a highly jointed rock mass below a concrete dam near heel.

- Dynamic settlement of earth or earth and rockfill dams should be less than 1 m or one-hundredth of the dam height, whichever is less. Longitudinal tensile strains (stress) should be less than permissible tensile strain (stress) in clay core. It is a tribute to the engineers that no failure of modern high earth and rockfill dams has taken place even during a major earthquake (~8 M) since 1975 all over the world, even when the earth is passing through an unprecedented peak of earthquakes. A filter upstream of a clay core is also provided to drain seepage water during rapid drawdown in many modern earth and rockfill dams. A recent trend is to build inspection galleries in the earth dams also like concrete dams due to better compaction equipment. The axis of a dam and bridge should be tilted with respect to the direction of tectonic movement as the anisotropic rock mass has minimum strength in the direction of tectonic movement. Abutments may be weak and prone to landslides in the direction of tectonic movements.

- The site should not be prone to formation of huge landslide dams (in unstable steep gorges subjected to large debris flow during long rains), causing severe flash floods upon breaking. An example is the temporary huge twin reservoir created by the landslide of the Pute hydroelectric station in Ecuador. Nor there should be deep-seated landslides near the dam site generating waves that would overtopple the dam as in a vajont dam. Further stability of villages above the reservoir rim is to be looked into for safety of the people.

- Approach tunnels, hill roads, and bridges should be planned and maintained well in landslide-prone mountains [subjected to heavy rains or cloud bursts (>500 mm per day) or rains for a long time] and snow avalanche-prone areas for easy construction of the project. Hill roads and vibrations due to blasting in openings should not damage houses upon the ground, if any.

- Planning of the underground infrastructure should be such that risk during construction is not high but risk is normal. As a result, the observed tangential strain ($\varepsilon_\theta = u_a/a$) in the crown should be less than the critical strain of rock mass ($\approx q_c/E_r$) [20]. The high lateral strain ratio ($\varepsilon_x/\varepsilon_z$) should not be too high in the cavern's walls (<0.85) [21]. Moreover, the convergence between well-supported walls should also be less than 1% of the width of caverns.

- Huge contingency funds should be made available for up to 30% of total cost for timely risk management in underground structures. Authorities should prepare risk management plans 1 and 2 in case of unforeseen geological hazards. Alternate huge diesel generators should be installed for an emergency power supply in openings. Pregrouting (before excavation of tunnels) is proving to be successful even for TBM.
- Extensive geodynamic monitoring of displacements and seepage into landslides along the reservoir rim and all caverns and reservoir-induced seismicity is undertaken along the entire hydroelectric project and connected to the Internet for global viewing using space imageries. The caverns are fitted with warning systems (connected to extensometers) for the safety of persons before and after commissioning of hydroelectric projects.

It needs to be kept in mind that the life of machines is about 35 years, whereas the life of underground openings is about 100 years. The state of the art prepared by Naidu [22] will help in the better maintenance of turbine blades damaged by silt. Further, the formation and instability of landslide dams may be monitored by space imageries, even across international boundaries.

It may also be mentioned that the temperature of underground openings stays at a level equal to the average temperature on the ground surface throughout the year. Thus usually air conditioning is not required, but if so, its cost is a bare minimum.

Because decisions made by planners and designers in the early stage of conception have a great impact on the cost of a project, they should study the current state-of-the-art of knowledge and experiences regarding exploration, design, contracting, bidding, and excavation using TBM. It is important that extensive geological exploration and geotechnical testing are done before detailed and final planning of a major hydroelectric project. Sharma [23] has highlighted rock engineering problems of hydroelectric projects in the Himalayas.

The following sections present only the geotechnical study of the underground powerhouse, as it is the most popular use of underground space. Besides, rock engineering problems are practically similar in all underground openings.

11.5 FUNDAMENTAL REQUIREMENTS

Underground construction usually involves working with brittle material of unknown strength and behavior. Hence, it is necessary to collect as much information as possible on the following characteristics:

- Geology, lithology, hydrology, and structural features for the purpose of classifying rocks from an engineering point of view
- Deformability
- Mechanical strength

- Original state of in situ stress
- Influence of excavation method on rock mass characteristics

The aforementioned information is normally collected during geological and geotechnical investigations. However, at many times, enough investigations are not carried out. Consequently, engineers run into innumerable difficulties during excavation.

11.6 PLANNING A CAVERN

Topographical constraints and many other engineering considerations sometimes make the planning process a complicated affair. It is worthwhile to look into the following points carefully:

- Availability of competent rock (a rock that is nearly or fully self-supporting) within the domain of the proposed underground structure.
- Solid rock cover equivalent to three times the width of the opening.
- Absence of any shear, joint, or fault within the proposed chamber space and in its near vicinity. The rock should thus be massive in nature.
- Preferably the alignment of the cavern should be nearly normal to the strike of rock beds and also parallel to major horizontal principal stress in a highly anisotropic stress field.
- Seepage of water should be expected within the rock cover area.
- The overburden pressure above penstocks or the bottom of caverns should be preferably more than the water pressure. To account for the stable rock slope of a hill (β), the Norwegian Geotechnical Institute recommends that the weight component of overburden (d) normal to the penstocks/HRT should be more than water pressure $\gamma_w \cdot h_w \cdot F$ [24], that is, $\gamma_d \cdot \cos\beta$ is more than $\gamma_w \cdot h_w \cdot F$, where h_w is maximum water head. The criterion is based on 45 case histories of unlined tunnels in Norway. The factor of safety F is about 1.10 [25]. Thus there are no chances of hydraulic fracturing near the top of penstocks due to a sudden gradient in seepage pressures, as overburden pressure is more than seepage water pressure [26].
- The width of a pillar between two adjoining caverns should be adequate. It should be preferably equal to one-half of the sum of widths of both the caverns or maximum height of adjoining openings, whichever is more, depending on the general quality of rock masses [7].
- A thick shear zone increases support pressure on the shotcrete in the adjoining rocks. Further, the concentration factor of tangential stress may be as high as nine in the cavern roof near the shear zone in the case of anisotropic rock mass, instead of nearly two for an isotropic rock mass [27]. The stress concentration at corners between the roof and walls may not be reduced by drilling a series of short holes there.
- In one of the powerhouse caverns in the Himalayas, it has been possible to monitor the roof convergence of 0.024 mm/month, which continued for

30 months due to separation of a weak layer above the roof. Then longer rock bolts were installed to stitch this weak layer with stronger rocks. Subsequently, it was observed that the roof convergence had stopped [28]. Deformations have been observed to stabilize near the shear zone also within a year.

- A steep rock slope orients the direction of principal stresses. The vertical displacement is not maximum near the crown of caverns nearby steep rock slopes. The vectors of displacements are slightly tilted accordingly so caverns (desilting and powerhouse) should be planned near a steep but stable rock slope.
- The effective thickness of the reinforced rock arch in the roof is nearly equal to the length of bolt or anchor (l_b) − spacing of bolts (s_b)/4 − fixed anchor length (FAL)/2 $\left(l_b - \frac{s_b}{4} - \frac{FAL}{2}\right)$. The same should be significant especially in caverns to provide effective support capacities, which are reduced in a water-charged rock mass [29].

It is normally not possible to meet all the points in practice; however, a location may be fixed judiciously, weighing the aforementioned factors carefully. For example, the orientation of an underground powerhouse transformer gallery has to be perpendicular to the power tunnels and penstocks. If the hill slope is not quite stable, the underground powerhouse should be located deeper than the unstable zone of rock mass.

11.7 DESIGN OF A CAVERN

The caverns are large in size and hence their design is a complex problem. The complexity is mainly due to site constraints and geological factors. The basic factors that influence the design philosophy are:

- Space and geometrics
- Geotechniques
- Orientation
- Excavation sequence and technique
- In situ stresses
- Earthquake forces
- Supports

A detailed, nonlinear, three-dimensional stress-cum-seepage analysis is done around the complex set of openings for the optimization of their cost and safety, considering folds, fault zones, and sequence of excavation. The following paragraphs discuss the aforementioned point briefly.

11.7.1 Space and Geometrics

Space requirements are dictated by the size of the electrical and hydromechanical equipment. The optimum requirements are to be worked out carefully, keeping in mind the minimum volume of excavation, least number of parallel openings ensuring expeditious construction, and least botheration during operation. A normal inverted "U"-shaped opening is preferred for the main machine hall chamber. Organization of the single opening or multiple openings should be such as to ensure the compactness of the system.

11.7.2 Geotechniques

Rock mass behavior and influence of geology on the same are the main considerations. Normally, competent rock masses are chosen that may require little or minimum external supporting barring normal PCC lining to ensure proper shape of the cavern. Not so good rocks are always available, hence a via media has to be struck between rock mass behavior and optimum support requirements.

A major problem is faced in determining the overall quality of rock masses in large underground openings. For example, excavation of an underground powerhouse starts with a drift in the crown. Supports are put on the basis of rock mass quality observed in this drift. Experience shows that with widening of caverns, zones of weaknesses such as thick shear zones, and soft intrusions are sometimes encountered that were never discovered in the drifts. Consequently, the postexcavation quality of a rock mass may be found to be too low. In fact, extra supports have been placed after full widening of caverns. Even in parallel system of caverns, one of them may give a misleading picture of good rock mass quality in the adjoining site for the cavern.

It is thus recommended that rock mass quality found from a drift should not be followed blindly for support designs in the full width of the cavern. The rock mass quality for full width of the cavern is sometimes reduced drastically to get the postexcavation rock mass quality for large underground caverns.

11.7.3 Orientation

The orientation of cavities vis-à-vis the strike of general rock beds and the relative disposition of two or more cavities with each other are points that need careful study. As already mentioned, the axis of the cavern should be across the strike of the beds, and the two or more openings should be separated by a distance preferably to equal the width or height of the largest cavern in the immediate vicinity. This criterion is most suitable under the majority of situations obtainable in the field.

11.7.4 Excavation Sequence and Techniques

The excavation sequence is planned carefully to ensure the least disturbance to the surrounding rock mass. It is appropriate to attack initially through a central draft and then advance toward the extremes. The rock mass vertically downward may be stripped out in lifts of manageable heights of not more than 3 m. Controlled or smooth blasting is recommended at the construction stage.

There may be towns or villages on the ground near a tunnel, cavern, or shaft. It must be ensured that peak particle velocity is less than the safe limit for the existing structures. As such, maximum charge per delay shall be controlled accordingly. The safety factor of the hill slope is checked, as ground subsidence is causing differential settlement of houses and so on due to repeated blasting.

11.7.5 In Situ Stresses

In situ stresses are normally known. Generally, vertical stresses are equal to the cover pressure, whereas horizontal stresses may vary. It is difficult to measure horizontal stresses in tectonically disturbed formations such as in the Himalayas. Knowledge of in situ stress is essential.

11.7.6 Earthquake Forces

There is no case of failure of a large underground opening during an earthquake. In tunnel openings also, failures are not known except perhaps in the neighborhood of active faults and portals. Even in natural caves located deep below the ground, collapses are not known. Explorers of caves have given accounts of their excitement while an earthquake occurred. There is a roaring noise at the entrance of the cave. They did not notice any vibrations or tremors. Although just on the surface, there was considerable damage to buildings and roads. It thus appears that most of the damage is done by surface Rayleigh waves, which are not too deep to affect the large underground openings, which are less than one-tenth of seismic wave lengths. In the Uttarkashi earthquake in India, no vibrations were experienced inside the Chhibro-Khodri underground powerhouse (see Section 11.9.2).

11.7.7 Supports

Supports may be conventional, that is, steel arches or rock bolts or cable anchors. Rock bolts or anchors with steel fiber-reinforced shotcrete (SFRS), followed by plain cement concrete lining or shotcrete, are generally appropriate under most circumstances. Desilting chambers are also lined with PCC lining.

Designing Steps

The designer, after acquiring the most needed knowledge, takes upon the actual design step by step as follows: dimensions are worked out on the basis of engineering requirements. However, most power plants may have a width between 15 and 35 m and a length varying between 200 and 500 m. Most rock masses may pose problems in the construction of such large cavities.

Openings of a span above 15 m are possible only in good and very good rock formations. Rock masses having a Bieniawaski's [30] geomechanical rating (RMR) above 70 may sustain openings above 15 m with little or moderate support. It is therefore advised to mark out an appropriate cavern width after determining the geomechanical rating. The top line in Figure 4.6 shows the upper limit of width of caverns according to rock mass quality (Q) [31].

In situations where rocks are not optimal, extensive supporting may be needed and such cases need great care. Figure 4.6 also shows the thickness of SFRS, spacing, and length of rock bolts. Foundations of turbines are designed to avoid their resonance using an elastic modulus of deformation of the underlying disturbed rock mass (E_e) rather than the modulus of deformation (E_d).

Estimating Support Pressure

Various classifications are currently in vogue and may be used for the preliminary support design. The rock mass classification of Barton and colleagues [31] may be used in most cases. It is worthwhile to map the major geological discontinuities that may traverse the proposed cavity. The geometry of the zones traced by these discontinuities should also be marked out to estimate the dead weight of blocks and wedges that may slide and consequently exert dead weight on the supports. The map of the discontinuities will also help in preparing plans for rock bolting and anchoring. According to Barton and co-workers [31], the roof support pressure is independent of the span of the cavern or tunnel in competent rock masses.

The vertical support pressure under most circumstances may be below 0.1 MPa. Due to the enormous height of the side walls, some lateral constraints may also have to be applied with the help of prestressed cable anchors where bedded and jointed rocks are encountered.

It is essential to undertake a finite element analysis of the cavern assuming the rock to be nonlinear. The ratio between horizontal and vertical stresses may be assumed according to measured in situ stresses. The analysis will give some idea on the areas of stress concentration, zones of tension and zones of high shear stresses, compressive stresses in pillars, and the possible mechanism of failure of the caverns. This study will be useful in designing the external support system. The colored graphics of displacements and stresses give an insight into the mechanics of interaction of underground openings with the rock slope, if any, and the effect of the sequence of excavations.

Designing the Sequence into the Mechanics of Excavation

Advance planning of the sequence of excavation is essential in planning the construction work. From the rock mechanics point of view, commencing the work through a central drift and then advancing laterally and vertically may be most appropriate. Excavation in the neighborhood of the extremes (outer boundary) can be done by smooth blasting so that the rock mass in the roof and walls is least disturbed. It may be useful to undertake a finite element method study to work out a most suitable and efficient sequence of excavation.

Samadhiya [27] and Al-Obaydi [32] have developed (ASARM) finite element Fortran language software for nonlinear, three-dimensional, anisotropic-coupled stress seepage analysis of three-dimensional openings in hydroelectric projects especially considering rock joints, faults, in situ stresses, rock mass strength, stress-dependent permeability, and sequence of excavation. Underground caverns may be divided into substructures for three-dimensional finite element analysis on a personal computer. Based on distinct element analysis, the 3DEC software of Itasca is alternative software for a blocky rock mass ($Q = 0.1$ to 100 and $H < 350Q^{1/3}$ m). The 3DEC accurately simulates a prestressing effect of intermediate principal stress along the axis of the opening. It also simulates realistically the high lateral strain ratio in the cavern walls in cases of critically oriented joints [21]. The other software perhaps does not consider complex, nonlinear anisotropy of a jointed rock mass.

The internal water pressure in HRT, TRT, surge shaft, penstocks, and desilting chamber may create a new groundwater table (GWT). The latter should be simulated in software. The worst condition is empty opening when PCC lining withstands compressive hoop stresses due to the GWT.

11.8 ADVANTAGES AND DISADVANTAGES

Advantages and disadvantages of hydroelectric projects are summarized in Table 11.2.

TABLE 11.2 Advantages and Disadvantages of Hydroelectric Projects

Serial No.	Advantages	Disadvantages
1	Hydroelectricity generation is the most hygienic and healthy compared to thermal power and nuclear power generating plants, as it generates minimum entropy or side effects.	High dams dislocate local people from ancestral villages and small towns due to submergence in the dam reservoir. Drying of a sacred river can hurt religious sentiments of the people. Safe discharge should be ensured. Tribal culture is damaged by workers from other places.
2	Water is a renewable source unlike coal, gas, and uranium. Unlike other sources, it causes minimum environmental hazards to nature as well as human beings.	Decommissioning of dam projects may create slight problems in the future.
3	Hydroelectricity is the least expensive and ideal for meeting peak demand of electrical power. Annual rate of inflation is about 1 to 2% per year against 10 to 15% for thermal and nuclear power due to costly inputs.	Construction period is too long, from 7 to 15 years, due to underground structures. A thermal power or gas project takes only 3 years. Transmission lines are long.

(Continued)

TABLE 11.2 Advantages and Disadvantages of Hydroelectric Projects—cont'd

Serial No.	Advantages	Disadvantages
4	Seepage of dam reservoirs into active faults and boundary of earth plates may reduce highest seismicity locally due to release of strain energy by reservoir-induced seismicity (RIS) slower with time.	RIS is increased but RIS does not exceed the natural maximum seismicity in that region. Underground buried channels can empty whole reservoir very quickly. Deep-seated rotational landslides near a dam can cause a high wave in a reservoir, which can damage a dam permanently.
5	Hydro power projects may also supply irrigation water to villages and drinking water to mega cities. Hydroelectric power projects generate local employment. Water-logged area is hardly 1 to 2% of irrigated land.	Dam traps silt supply that is otherwise supplied to plains for soil fertility. A reduction in river flow downstream of a dam reduces the quality of river water. Global warming and receding of glaciers may reduce discharge in river and reduce electricity generation in the long term.
6	Hydroelectric power projects help in controlling floods in heavy rainy seasons in case of high dams.	In landslide-prone deep gorges, landslide dam reservoirs can flood a tail race tunnel and underground powerhouse machines frequently after their failures. Ancient monuments have to be shifted away from the reservoir.
7	Underground structures are safe from all natural disasters and nuclear attacks and do not occupy vast surface area. Excavated debris may be used to widen terraces in villages. Excavated rocks may be used as building materials freely.	Completion of project is invariably delayed due to tunneling hazards in long tunnels and site-specific problems. Construction of surface structures on a stable mountain is fast and economical due to easy access.
8	There is no fly ash and radioactive nuclear waste.	However, surface structures are unsafe during wars and in landslide-prone areas.
9	Hydroelectric projects develop the infrastructure and economy of poor hill people, not only mega cities. They have many life-giving healthy benefits (see No. 5). No failure of high earth dams has occurred since 1970, even during major earthquakes all over the world.	Dam burst and cracks may cause immense damage to life and property downstream.

TABLE 11.2 Advantages and Disadvantages of Hydroelectric Projects—cont'd		
Serial No.	Advantages	Disadvantages
10	Burning of millions of tons of coal is avoided, unlike thermal poor plants. Availability of electricity from hydroelectric projects, in fact, checks the demand of hill villages to cut forests for fuel and fire wood.	Thermal and nuclear power generating plants are high entropy, whereas hydroelectric projects are the least entropy technology with least unhealthy side effects.
11	Reservoir lakes are recreational spots for boating and enjoying nature and are a life support system of flora and fauna in its vicinity.	Forests are submerged and the ecosystem is damaged by a large dam reservoir. Reservoirs can cause deep-seated rotational landslides. Water-borne diseases may increase around stagnant water reservoirs.

Dislocated families should enjoy a better quality of life and education at the place of resettlement than they enjoyed at the original habitat. Further, in all new dams, outlets should be provided to ensure at least 10% flow downstream as a general policy for ecoregeneration. The cumulative effect of all dams along a river ecosystem should be examined by a central river valley authority.

11.9 CASE HISTORIES

Three selected educative case histories are presented here for enriching the knowledge of engineers and geologists.

11.9.1 Churchill Falls Hydroelectric Project

The development of hydroelectric potential at Churchill Falls (5225 MW) in Montreal, Canada, was, without doubt, one of the outstanding engineering challenges in the 1970s. It has one of the largest underground powerhouses with eleven 475 MW generators of large size with a water head of 310 m. Penstocks were inclined at 55° to allow self-mucking during excavation. The tail race tunnel is approximately 1.6 km long. The powerhouse is 24 m wide and 292 m long. Benson and colleagues [6] have given an educative case history of this pioneering hydroelectric project (Fig. 11.1).

The rocks are gneisses with bands of diorite and some shear zones (<30 cm). The rock mass quality Q = 10 (q_c = 107 MPa, E_d = 6–20 GPa, RQD = 94%, four joint sets, fracture frequency = 1.5 to 3 per meter, very low absorption of water during test). The rock mass is rated good and suitable for a large underground powerhouse. The most economic arrangement of caverns is where rock pillars are as small as possible. The shapes of openings are governed by the smooth flow of horizontal principal stress trajectories around all the adjacent openings. As such, plane strain, elastic, two-dimensional stress analysis was done by the finite element method. The ratio of horizontal to vertical stress was found to be about 1.7 (1.1 to 1.9). Parametric analysis yielded the location and extent of tensile stresses above the roof, walls, and pillars of all caverns. The rock bolt system with a chain-link mesh was designed to extend beyond these tensile (or stress relaxation) zones. The pattern rock bolting followed immediately after excavation to the main arch in the soundest condition possible according to the New Austrian Tunnelling method. The chain link provided safety to workers below. The four layouts were analyzed before selecting a final design. The safe compressive strength of the pillars was assumed to be 33 MPa. The circular shape of the arch of the roof is very important to minimize the tensile stress zone without an excessive cost of extra excavation. The lengths of rock bolts are one-third to one-fourth times that of the span in the arch and one-fourth to one-fifth times that of the height of the walls. The rock bolts were inclined at 20° with horizontal to be nearly perpendicular to the planes of foliations on the walls. Benson and co-workers [6] highlighted the engineering practice that the ratio between rock bolt length and width of opening decreases with increasing size of the caverns.

The measured displacements never exceeded 10 mm in the arches of openings. The recorded roof displacements were one to three times the theoretically predicted values by the finite element method. The measured slips along weak shear zones were negligible. The maximum measured wall displacement was about 15 mm. The mortality rate of instruments was barely 5%. Rock bolt load cells were also provided. Discontinuous inelastic displacements were also observed at some locations. Extra rock bolts were installed there.

It was considered important to maintain a free draining rock mass around major openings to prevent the buildup of seepage pressures. As such, drainage holes of 1.5 m depth were drilled at a spacing of 4.5 m both in the roof and in the walls.

Extensive contact and consolidation grouting was done around steel liners of the penstocks. The interaction analysis of rock mass–concrete steel liner was done by the plane strain, two-dimensional finite element method considering the elastoplastic behavior of the jointed rock mass. Piezometers were installed behind steel liners to monitor seepage pressures. The concrete lining was assumed to be radially cracked due to hoop tensile stresses. The rock mass was assumed to be radially cracked up to the end of the tension zone of 5 m thickness.

Cooperation between the contractor and engineers was the cause of success. Finally, Benson and colleagues [6] recommended that rock mechanics and engineering geology experts should not only be part of the group but also be involved in the construction work of the hydroelectric projects.

11.9.2 Chhibro Underground Powerhouse

The Yamuna Hydroelectric Scheme Stage-II in India harnesses the hydropower potential of the river Tons, which is a tributary of the river Yamuna. The available head of 188 m is being utilized in two stages. Stage-I utilizes the head of about 124 m along the first river loop between Ichari and Chhibro to generate 240 MW of power. To avoid large-scale excavation of steep slopes, the powerhouse chamber is located underground. Its size is 18.2 m wide, 32.5 m high, and 113.2 m long. This cavern is excavated in a band of limestone of 193 × 217 m horizontal extent. A major shear zone passes within 10 m of the lowest draft tube level in the powerhouse area. It is perhaps the first experience of construction of a large underground powerhouse in weak rock masses in 1972. Mitra [10] had studied this project extensively.

The powerhouse cavity has a span of 18.2 m and the height is about 32.5 meters from the lowest excavation level up to the crown. The powerhouse was excavated through limestones of Mandhali Series. The engineering properties of the rock material are given here:

- Deformation modulus = 5.6 GPa
- Unconfined compressive strength of rock material = 50 MPa
- Flexural strength of rock material = 18 MPa
- Shear strength parameters, Φ = 34 and c = 17.6 MPa

The support pressure for the computation of support dimensions had been assumed to be equivalent to the dead weight of a rectangular wedge having a height of 8 m (according to Terzaghi classification). This gave a rock pressure of about 0.2 MPa.

Support Details

For immediate supports, shotcreting was done to prevent popping. The crown was provided with a steel arch as a permanent support. The high sides of the cavity were strengthened by systematic cable anchoring and by prestressing them. Figure 11.5 gives the section of the prestressed cables.

Vertical walls were supported by prestressed anchors and shotcreting with chain link fabric as per the following sequence:

1. Drilling of holes to a predetermined angle of a predetermined diameter by use of a guide tube before taking up excavation in a bigger cavity
2. Pregrouting the rock strata using the holes drilled as just mentioned
3. Redrilling the holes after pregrouting
4. Installation of cables
5. Stressing the cables against the anchorages
6. Restressing the cables after the desired period
7. Final grouting of the hole through holes drilled in the anchor plates
8. Cutting off extra wires and covering them with shotcrete

The system used for this work is called the Gifferd Udall CCL prestressing system "known as a single wire system." The anchors provided were of 45 to 60 tons capacity. Spacing of the anchors was such that the dispersion of forces from the bearing plate could hold the rock mass, as shown in Figure 11.5. The twin tunnels acted as drainage galleries to keep the roof of the cavern dry and were used to prestress these cable anchors.

Monitoring Performance of Caverns

Instrumentation of large underground caverns is very important. The objectives of such instrumentation should be:

- To gather information on overall elastic modulus of the rock mass, which should then be checked with the assumed value in finite element analysis
- To locate unstable wedges or zones of rock masses within the roof or the side walls
- To study the performance of support systems
- To trigger a warning system when the opening tends to become unstable

Several borehole extensometers have been installed at the Chhibro underground powerhouse. At one place, an extensometer is connected to an electric bell, which will be switched on when displacement exceeds the permissible limit (of 5 mm). Vibrating wire strain gauges have been fixed on the bottom flanges of steel ribs in the crown of two caverns. The increase in support pressure is thus being monitored. So far caverns are performing well and there is no danger of collapse.

Effect of Earthquake

An earthquake of 6.3 magnitude occurred on October 21, 1991, which was centered near Uttarkashi, 100 km away from the project site. The earthquake devastated the entire Uttarkashi and about 100 km away from the project site. Recorded damage in the Chhibro powerhouse cavern on account of this earthquake was limited to minor cracks in the region closest to the shear fault zones. The damage, as described by Mitra and Singh [33], is as follows:

- Out of the eight extensometers installed on the side walls of the powerhouse, only two on the downstream wall adjacent to the control room (nearest to the underlying shear fault zone) recorded any significant rock deformation. These deformations were of the order of 1 to 4 mm. In addition, a deep crack of 2 to 4 mm width formed diagonally up to a length of 3.5 m between these two extensometers.
- Horizontal hairline cracks were observed on each column of the control room and the downstream side wall at heights of 0.5 to 2.5 m.
- Two horizontal 1-mm-wide cracks, of lengths about 5 m, were found in the portal at the main entrance of the powerhouse adit.

- Two vertical cracks of 0.5 and 4 mm width with a spacing of about 80 m were observed inside the adit at a height of about 1 m. However, these cracks appeared to have formed in the shotcrete lining.
- Anchor plates supporting prestressed rock anchors in the expansion chamber adit appeared to have stretched slightly and may have caused lining cracks.

There was no damage in the sections of the powerhouse complex away (up to a distance of width of opening B) from the shear/fault zone. An analysis by Mitra and Singh [33] shows that the dynamic support pressures are negligible compared to the long-term support pressures in the roof of the chamber near the shear fault zone due to residual strains in the nearby rock mass. Repairs have been made and the cavern is stable.

The aforementioned study shows that seismographs should be installed inside a tunnel across active faults to record seismic peak acceleration in the roof, walls, and base. Further, caverns should also be safe even in weak rock masses during an earthquake.

11.9.3 Tala Hydroelectric Project, Bhutan

It is an engineering miracle and educative project with the following features [34]:

- A concrete gravity dam 92 m high and 130 m long at the top located at Wangkha about 3 km downstream of the Chukha powerhouse
- Three underground desilting chambers of $250 \times 13.9 \times 18.5$ m size for the removal of suspended sediments of 0.2 mm size and above
- Head race tunnel of 6.8 m diameter and 23 km length
- Surge shaft 15/12 m diameter and 184 m in height
- Two pressure shafts of 4 m diameter and 1.1 km long each trifurcating into penstocks of 2.3 m diameter
- Machine hall cavern of $206 \times 20.4 \times 44.5$ m to house six Pelton turbo generators of 170 MW capacity each (1020 MW)
- Transformer hall cavern $190 \times 16 \times 26.5$ m to accommodate 19, 70 MVA 13.8/400-kV transformers
- Tail race tunnel of 7.75 m diameter and 3.1 km length
- Two double circuit 400-kV transmission lines from the Tala powerhouse to the Indo-Bhutan border of 140 km circuit length
- A 400-kV/220-kV, 200 MVA capacity interconnecting substation at Malbase, Bhutan
- Construction on a fast track between 1998 and 2007 due to good management conditions

Figure 11.6 shows the location of dam, diversion tunnel, and three desilting chambers with a silt flushing tunnel.

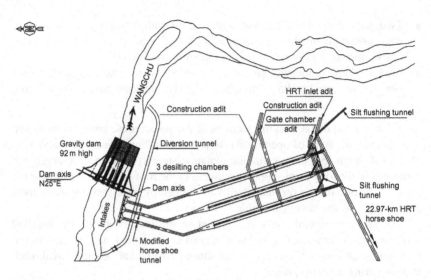

FIGURE 11.6 Layout of dam complex and desilting chambers.

The head race tunnel passed through quartzite and biotite schist and was aligned along strikes of foliation (Fig. 11.6). Steel fiber-reinforced shotcrete, together with rock bolts, was used. The rebound of SFRS was 4 to 9%. At one location, adverse geological conditions were met, and rock mass had an unconfined compressive strength (UCS) less than 1 MPa. Consolidation grouting and grouting by polyurethane were tried in the original tunnel, yet the tunnel supports collapsed, so the tunnel alignment was diverted locally by passing the tunnel where they got slightly better rock conditions. The thick shear zone was again met, and a severe squeezing ground condition was again faced due to high in situ stresses. So heavy steel ribs (of capacity 150 T at spacing of about 60 cm) were embedded in thick SFRS, and invert struts were welded to the steel ribs in this adverse geological reach. The maximum observed load from load cells in rib was 170 T. Reinforcement by a forepole umbrella of 21 rock bolts of 9 m length was provided. The tunnel face was supported by SFRS, and a drainage drill hole was made to release seepage pressures. The wall convergence was observed to be 17 cm. Upheaval of the tunnel bottom due to squeezing was recorded as 4.8–5.3 cm. The weak rock mass was excavated by scooping [35].

Project commissioning could have been achieved in a lesser time frame but there was a delay due to the most unprecedented natural calamity of August 2000 when 1700 mm of rainfall was recorded in just 3 days. Kilometers of freshly constructed roads vanished, as if they had not existed at all. In that deluge, precious lives were also lost in landslides and flowing streams. About 15,000 persons and 84 engineers were working during the peak time. A good number of skilled workers left the worksites in panic.

Restoration of washed-away access roads and replacement of equipment and bridges damaged in the storm took more than 6 months. Within a few months of resuming the project, nature was unkind once again, and in May 2001 almost one-third of the crown length of the powerhouse cavern collapsed—restoration of which took 10 months, up to April 2002. Construction of pressure shafts posed serious problems as five large cavities were formed due to rock falls into the already constructed shafts, resulting in multiple blockages. Removal of the blockages and treatment of cavities took another 12 months. Rock mechanics studies have been carried out. The entire contract for in situ instrumentation was given to two firms. The cost of instrumentation was about 3% of the total project cost. The observed large displacements in the right and left walls of the machine hall cavern are 13 and 23 cm, respectively. Yet the cavern appears to be stable. The software 3DEC predicted a wall convergence of 17 cm against the observed value of 36 cm.

Under afforestation plans, more than 725,000 saplings were planted at various locations, of which more than 500,000 have survived and have grown fast.

Machine Hall Cavern

The machine hall cavern is 206 m long, 20.4 m wide, and 44.5 m high and has a rock cover of approximately 400 m [34]. The pillar width with adjoining transformer hall is 39 m [35]. The longitudinal axis of the machine hall cavern has a trend of N37°W. Geological formations at the machine hall location consist of a fresh and hard bedded sequence of quartzite, quartzite and amphibolites schist partings. These rocks are highly puckered and folded into tight synform and antiform. The general foliation trend varies from N65°E-S65°W to N70°W-S70°E with 35° to 60° N25°W to N20°E dips. The RMR varies from 19 to 50 and GSI from 27 to 52, and the rock mass quality (Q) is in the range of 0.11 to 14. It classifies the rock mass as fair to good.

The in situ stresses at the site had been estimated to be as follows: vertical stress = 10.9 MPa (from overburden); minimum horizontal stress = 9.5 MPa (approximately normal to cavern axis); and maximum horizontal stress = 14.2 MPa (approximately parallel to the cavern axis). The UCS of the intact rock is reported to be 63 MPa.

The support system was composed of 12-m-long and 32-mm-diameter fully grouted-tensioned threaded (Dywidag) rock bolts with 200-mm-thick shotcrete. The observed tensile strength of steel was 1020 MPa with elongation of 4 to 11% and of capacity 57 T. Later on, instrumentation was done to monitor the response of the rock mass during the excavation and postexcavation period [34]. A major rock fall is reported to have occurred on May 7, 2001 between reduced distance (RD) 55 m to RD 63 m (Q = 7.5) in the crown portion exposing rock bolts [34]. The height of the cavity formed was approximately 55 m. Another massive rock fall was reported on May 29, 2001 between RD 135 m to RD 150 m (Q = 7.5–12), which extended on both sides subsequently on May 30 and June 2, 2001. Rock bolts in this portion were hanging with the end anchorage intact and lower end of most of the bolts broken. The failure of rock bolts and probable causes of collapse have been analyzed by Chowdhury and Singh and Sthapak [34]. The failure of rock bolts was reported to have followed the following pattern:

- Initially, the plate and end anchorage was intact; the rock in between gave way. The plates were bent.
- The end anchorage was intact; the bottom plate got sheared off. Dilation of the rock mass might have occurred at a higher level and the rock bolt could not support the support pressure and failed at the plate junction.

Singh and Sthapak [34] have mentioned a few probable causes of collapse. One of the main causes has been attributed to dilation and stress release in the rock. After this failure, the 15-m-long rock bolts of 57 T capacity and at a spacing of 1.5 m center to center have been designed to stabilize walls using the software 3DEC [35]. The thickness of SFRS is 150 mm. Conventional steel ribs of ISMB 350 at a spacing of 0.6 m have been used in roofs to prevent its failure again. Singh and Sthapak [34] have reported observations from an instrumentation program followed at later dates in the machine hall. Eight instrumented rock bolts were installed in the machine hall. A variation of loads in 12-m-long instrumented rock bolts on the walls of the machine hall with time (September 2002 to April 2007) has been reported. The load in the rock bolts reached a maximum of 43.01 tonne (430.1 kN), after which the bolt failed and the load fell down to 1.58 tonne (15.8 kN). A maximum of 353.35 mm of convergence ($\approx$1.77% of the width of machine hall) has been reported for the side walls. Most of the convergence took place during excavation and very small convergence occurred during the postexcavation period. The longitudinal axis of the machine hall trends at N37°W. The joint set J4 is also reported to be striking in a N50°W to N30°W direction. Joints belonging to set J4 therefore strike parallel to the longitudinal axis of the cavern. The dip of these joints varies from 60 to 70°. The friction angle of the joints is reported to be 25°. Thus critical orientation of the joint is $\theta \approx 45° + \Phi_j/2 = 57.5°$.

The high value of wall deformation of the machine hall ($\approx$1.77 %), failure of instrumented rock bolts in the wall, and crown collapse at the machine hall in the Tala hydroelectric project are some of the manifestations of high lateral (horizontal) strains due to critical orientation of joints striking nearly parallel to the axis of the opening at the site [35].

It may be concluded that the critically oriented joints striking parallel to the longitudinal axis of the underground opening may slip permanently due to stress release and also create voids that result in high lateral strain in the mass. The tendency to create voids is maximum if the joints dip at a critical orientation. A high lateral strain may result in excessive wall closure and failure of rock bolts. It is, therefore, advisable that the time of installation of bolts be delayed up to stand-up time according to the unsupported span of the rock mass. The slip along joints may be arrested by rock bolts of adequate capacity in the reinforced rock mass. In case rock bolts are not able to prevent slip along joints, the maximum lateral strain in the rock walls must not exceed the yield strain of the steel bolts for the stability of rock bolts. Thus resingrouted steel bolts of a high yield strain should be used rather than brittle bolts of high tensile strength in the aforementioned geological situations. Fears are expressed that resonance (high amplitude of peak ground acceleration) of generator foundation may take place as observed vibrations are quite high.

After Tala

The best known natural resource of Bhutan is water, flowing down the steep slopes of its rivers, with an estimated hydropower potential of 30,000 MW. The Sankosh earth and rockfill dam in Bhutan would produce 4000 MW electricity and would have a bigger reservoir than the Tehri dam in India. The most dependable market for this commodity across its border is India, where the installed capacity of 130,000 MW is short by 22,000 MW of hydropower for an ideal 40:60 of hydro:thermal mix of the power system.

11.10 SUMMARY

There are many uses of large underground openings, limited perhaps only by one's imagination. Their most important application is in the construction of underground power stations. The modern trend for the planning of hydroelectric projects is toward making long power tunnels and deep penstocks, with air bubble surge chambers, if possible. In this way, hydraulic head available to a power station becomes enormous. The geotechnical problems created by large underground openings are challenging. However, enough experience has been gained in solving these problems. This has been possible as rock pressures are practically independent of size so there is no need of very heavy supports. The sequence of excavation is an important decision. Here finite/distinct element analysis is generally helpful. Finally, it costs nothing for monitoring rock movements in a cavern during various phases of excavation and support. Quality control is essential in weak rock masses. In the end, three case histories of underground powerhouses were given to illustrate various design considerations. The Q system is found to be the best for tunnels and caverns except in adverse geological conditions when compared to RMR [36].

The cost of hydropower is about half the cost of nuclear power. Therefore, investors in hydropower projects are likely to get a good return in the long term due to the low cost of power generation and the many life-giving benefits as compared to any other power projects. The inflation rate of thermal and atomic power is more than 10 to 15% per year due to costly inputs, whereas the inflation rate for hydropower is barely 1 to 2% per year.

REFERENCES

[1] Hudson JA. Rock engineering systems: theory and practice. London: Ellis Horwood Ltd.; 1992. p. 185.

[2] Vansant C. View point—hydro's efficient advantage. Hydro Rev Worldwide (HRW) 1994;2(1):8.

[3] Barnes MJ, Beggs SL. Hydro in Asia and Pacific: strength in diversity. Hydro Rev Worldwide (HRW) 1994;2(1):10–18.

[4] Broch E. Why did the hydropower industry go underground? Norwegian Tunnel Soc Pub 2006;15:131–8.

[5] Jaeger C. Rock mechanics and engineering, chapters 10 and 12. Cambridge University Press; Cambridge, 1972.

[6] Benson RP, Conlon RJ, Merrittee AH, Joli-coeur P, Deere DU. Rock mechanics of Churchill Falls. ASCE symp underground rock chambers, Arizona; Phoenix, ASCE, Mtg. Water Resources Engng, 1971. p. 407–86.

[7] Hoek E. Practical rock engineering, chapter 6, 2000 ed.; 2000 (www.rocscience.com/hoek/practicalrockengineering.asp) Accessed in 2008.

[8] Christensen JB. Unlined compressed air surge chamber for 24 atmospheres pressure Jukla power plant. In: Balkema AA, editor. ISRM symp rock mechanics: caverns and pressure shafts, vol. 2. Rotterdam; Aachen, Germany: International Society for Rock Mechanics; 1982. p. 889–92.

[9] Dolcetta M. Problems with large underground stations in Italy. ASCE symp underground rock chambers, Phoenix Arizona; 1971. p. 243–86.

[10] Mitra S. Studies on long-term behaviour of underground power-house cavities in soft rocks. Ph.D. thesis, civil engineering department, IIT Roorkee, India; 1991.

[11] Mitra S, Singh B. Long-term behaviour of a large cavern in seismically active region of lesser Himalaya. Eight ISRM Congress on rock mechanics. Tokyo, Japan; ISRM, A.A. Balkema; 1995. p. 1295–8.

[12] Verman M. Rock mass: tunnel support interaction analysis. Ph.D. thesis, civil engineering department, IIT Roorkee, India; 1993. p. 267.

[13] Mehrotra VK. Estimation of engineering parameters of rock mass. Ph.D. thesis, civil engineering department, IIT Roorkee, India; 1992. p. 267.

[14] Swamee PK. Design of sediment flushing outlets. Int J Sediment Res China 2002;17(4):314–22.

[15] Asthana BN. Personal communication with Bhawani Singh. IIT Roorkee, India; 2007.

[16] Helwig PC. Personal communications with Bhawani Singh. IIT Roorkee, India; 2007.

[17] Barton N. Q_{slope} method, training course on rock engineering. New Delhi: CSMRS; 2008. p. 343–63.

[18] Swamee PK. Reservoir capacity depletion on account of sedimentation. Int J Sediment Res China 2001;16(3):408–15.

[19] Krishna J. Seismic zoning maps in India. Curr Sci India 1992;62(122):17–23.

[20] Singh M, Singh B, Choudhari J. Critical strain and squeezing of rock mass in tunnels. Tunnel Undergr Space Technol 2007;22:343–50.

[21] Singh M, Singh B. High lateral strain ratio in jointed rock masses. Eng Geol 2008;98:75–85.

[22] Naidu BSK. Silting problems in hydropower plants—a comprehensive overview, 2nd international conference on silting problems in hydro power plants. Editors SP Kaushis and BSK Naidu, CBIP, New Delhi, India; 2002.

[23] Sharma VM. Geotechnical problems of water resources development in India: the role of instrumentation. 32nd IGS lecture, IIT Bombay, Mumbai; 2010. p. 1–93 [republished in Indian Geotech J 2011;41(1):1–47].

[24] Buen B, Palmstrom A. Design and supervision of unlined hydropower shafts and tunnels with head up to 590 metres. ISRM symp rock mechanics: caverns and pressure shafts. International Society for Rock Mechanics, ISRM International Symposium, Aachen, Germany; 1982;2:567–74.

[25] Helwig PC, Besaw DM. CAT arm power tunnel geology and underground works. Canadian Electrical Association Hydraulic Power Station and Cancold, Newfoundland; 1986. p. 23.

[26] Barton N. Deformation phenomenon in jointed rocks. Geotechnique 1986;36(2):147–63.

[27] Samadhiya NK. Influence of anisotropy and shear zones on stability of caverns. Ph.D. thesis, civil engineering department, IIT Roorkee, India; 1998. p. 334.

[28] Goel RK. Status of tunnelling and underground construction activities and technologies in India. Tunnel Undergr Space Technol 2001;16:63–75.

[29] Singh B, Viladkar MN, Samadhiya NK, Sandeep KG. A semi-empirical method of the design of support systems in underground openings. J. Tunnel Undergr Space Technol 1995;3:375–83.

[30] Bieniawski ZT. Case-studies-prediction of rock mass behaviour by the geomechanical classification. Second Australia-New Zealand conference on geomechanics, Brisbane; 1975. p. 36–41.

[31] Barton N, Line R, Lunde J. Engineering classification of rock masses for the design of tunnel support. Rock mechanics. New York: Springer-Verlag 1974;6:189–236.

[32] Al-Oaydi M. Stability of underground structures in anisotropic rock masses. Ph.D. thesis, civil engineering department, IIT Roorkee, India; 2006. p. 497.

[33] Mitra S, Singh B. Influence of geological features on long-term behaviour of underground powerhouse cavities in lower Himalayan region: a case study. J Rock Mech Tunnel Technol India 1997;3(1):23–76.

[34] Singh R, Sthapak AK. International workshop on experiences gained in design and construction of Tala hydro-electric project Bhutan. Organised by ISRMTT, CSMRS, THAPGB; 2007. p. 347 (see papers of R.N. Khazanchi, p. 1–11, A.K. Chowdhary, p. 93–105, Sripad et al., p. 269–282 and see also p. 23 and p. 333).

[35] Singh R. Excavation and support system for power house cavern at Tala hydro-electric project in Bhutan. J Rock Mech Tunnel Technol 2008;14(1):11–22.

[36] Singh R. Tunnelling in extremely weak rock mass conditions at Tala hydroelectric project. Bhutan, lecture on 19.6.2009 at IIT Roorkee, India; 2009.

Chapter 12

Underground Shelters for Wartime

Except our own thoughts, there is nothing absolutely in our power.

Rene Descartes

12.1 GENERAL

Natural and man-made underground facilities have played an important role in warfare and national security for more than 5000 years. Underground chambers were used for hiding places and escape routes in Mesopotamia and Egypt from 3500 to 3000 B.C. and continue to play an important role in the ongoing conflict in Afghanistan. Some notable 20th-century uses of underground facilities for warfare and national security include dozens of underground factories constructed beneath Germany during World War II; the Cheyenne Mountain Operations Center in Colorado; as many as 1000 underground facilities estimated to exist beneath the Korean demilitarized zone; and an extensive network in Vietnam and countless natural and man-made caves used by Al-Qaeda forces in Afghanistan. Large man-made cavities in salt domes along the Gulf Coast, some of them larger than 17 million cubic meters in volume, are used for the U.S. Strategic Petroleum Reserve. The details of underground facilities used for military or national security purposes are classified, but there is no reason to assume that they are not on the scale of underground civil projects. The largest cavern ever constructed in rock was a 61-m (200-ft)-wide hockey arena constructed for the 1994 Winter Olympics in Norway, and underground mines in many parts of the world consist of smaller passages that extend for many miles [1].

Extensive underground facilities have also been constructed to maintain communications and house the U.S. government in the event of an attack. An underground facility known as "Site R" exists within Raven Mountain, Pennsylvania, and is thought to have been the location from which Vice President Cheney and other officials worked in the aftermath of the September 11, 2001, terrorist attacks. Construction of "Site R" was authorized by President Truman and completed during the early 1950s. Declassified information dating from the construction period describes a three-story underground facility with more than 18,000 square meters of floor space and room for more than 5000 people. The existence of another extensive underground facility beneath the Greenbrier Resort in West Virginia, constructed to house the U.S. Congress in the event of a nuclear attack, was made public in 1992 [1]. Safety costs money.

Although they can be expensive and difficult to construct, underground facilities offer two important advantages over surface structures. First, they are almost completely hidden from view and activities within them can be invisible to even the most sophisticated intelligence satellites. Second, their overburden depth can make them resistant to conventional and some nuclear attacks. Additional advantages include lower long-term maintenance costs (because underground structures are not exposed to weather) and lower heating and cooling costs (because temperature is constant in underground environments). The detection and characterization of underground facilities and the development of technologies to defeat hardened underground facilities are among the principal goals of modern military geologists.

One issue important for military- or security-related underground facilities, but generally not for civil structures, is their vulnerability to attack by conventional or nuclear weapons. The vulnerability of an underground facility to a conventional weapon attack is a function of its depth, the strength of the overlying rock, and the penetrability of the soil or rock exposed on the earth's surface above the facility. Knowledge of these properties is essential to those designing facilities to survive attacks, as well as to those designing specialized earth penetrating weapons. The geologic information necessary to evaluate the vulnerability of a facility has been given the name "strategic geologic intelligence" by some military geologists. Few, if any, underground facilities can withstand a direct nuclear attack. Let us prepare ourselves for underground space technology wars.

12.2 STATE-OF-THE-ART DEFENSE SHELTERS

The concept of engineered structures to protect against artillery can be said to have originated in the late 17th century with the Marquis de Vaubain. Field fortifications using overhead protection against high-trajectory artillery fire were used to some extent in the American Civil War. In World War I, the intense use of artillery and the beginning of aerial bombardment led to the extensive use of field fortifications with massive overhead protection. Development between the wars resulted in the refinement of modern protective construction against modern conventional weapons. This technology made extensive use of armor plate and reinforced concrete, which was developed late in the 19th century. The design procedures for reinforced concrete also underwent a great deal of development between the world wars [2].

Well-known examples of this technology were the Maginot Line between France and Germany and the Fort of Eban Emael on the Belgian frontier. Although these fortifications were outflanked and neutralized by brilliant unconventional tactics and audacity in the 1940 German Blitzkreig, their ability to protect their inhabitants against bombardment was never challenged. Indeed, the manifest strength of the fortifications effectively deterred frontal assault.

Shelters to protect civilians against blast and fragments became highly developed in World War II. Generally known as "bomb shelters," the structures included blast doors, protection against chemical agents, and the use of subsurface construction. Germany in particular developed some very effective technology against aerial weapons because it was subjected to very heavy bombing (U.S. Strategic Bombing Survey, 1945, 1947 [2]). Notably their designs for civilian shelters ("bunkers") and their construction of massive concrete protective structures for their submarine were effective.

The German "*Sonnenbunkers*" were aboveground reinforced concrete buildings designed to resist direct hits from aerial bombs (originally 250-kg bombs and, eventually, 1000-kg and larger bombs). These shelters were expensive, requiring up to 3 m^3 of concrete per occupant in the smallest (500-occupant) size and 1.8 m^3 of concrete in the larger size (4000–4800 occupants), even when crowded to 5 people/m^2.

It was decided early that bunkers could not be afforded for the entire population but would be restricted to 5% of the population of some 70 cities designated as strategic targets; however, bunker space was extended for about 15% of the population of those cities by 1944. The rest of the population had reinforced basements or belowground tunnel shelters.

Few improvements had to be added to the World War II bomb shelters to make them effective against nuclear weapons. The massive concrete and underground construction provided inherent protection against nuclear radiation. High-performance shelters against the effects of nuclear weapons required the addition of radiation protection of the entrances, either through the use of more massive doors or entryways incorporating one or more turns. The long duration of the blast pressure from large nuclear weapons required the addition of shock isolation inside the shelters to protect against ground motion and the insertion of blast valves in the ventilation air intakes to prevent the shelters from being filled with high-pressure air. Filters were generally included to keep out radioactive dust and fallout. The conduction of a heat shock wave from a mega nuclear bomb may make survival difficult. Therefore, a minimum overburden of 300 m is considered safe for caverns from surface nuclear attacks in modern times.

Tests of shelters against low- and intermediate-range yields (tens to hundreds of kilotons) at more than 1 atm overpressure quickly showed that initial nuclear radiation was an important design parameter, requiring several meters of earth cover on the shelters for shielding.

Coincident with the cessation of nuclear testing in the early 1960s, technical developments in shelters for civilians were directed at ways of reducing the costs of a shelter rather than at improving the protection it could offer. Researchers accepted that it was not cost-effective to seek shelter designs for the civilian population that could survive a very close detonation of nuclear weapons.

With the discovery of fallout from the testing of large-yield weapons, it was recognized that any system of protection would require that fallout radiation protection be provided to the entire population outside of the target areas.

"Slanting" began to be explored in the 1960s as a means of reducing the cost of shelters for civilians. This technique consists of slightly modifying construction intended primarily for other purposes in such a way that protection against nuclear effects is developed. Basements, subways, and tunnels being built for other purposes are candidates for slanting.

Explorations of the potential of "best available" shelters were carried out in the 1960s and 1970s. It was discovered that the basements of buildings, especially those with reinforced concrete first floors, could provide significant protection (0.8 atm—12 psi or more) against blast, as well as radiation, without any modification. People would have to stay out of entryways, where high-velocity winds could propel them with lethal velocities against the floor or wall.

These discoveries led logically to the concept of "upgrading," in which suitably constructed floors are reinforced with columns that can be moved during a crisis. Additional shielding in the form of earth can also be added to the first floor and piled against exposed walls. Given time, the protective capability of a structure not specifically designed as a shelter can be improved considerably in this way.

The concept of an "expedient shelter" showed substantial development in the 1970s. This term has been adopted to mean a shelter that is constructed during a crisis from materials, resources, tools, and labor available. Most expedient shelter designs are covered trenches with either shored or unshored earth walls. Aboveground and semiburied versions can be used in regions characterized by a high water table. Most expedient shelters are constructed from wood, and all are covered with earth to provide significant radiation protection. The covered trench versions usually provide "fallout protection factors" above 200. Some versions of expedient shelters have demonstrated survival of blast overpressures in excess of 7 atm (100 psi or 0.7 MPa).

A "protection factor" (PF) or "fallout protection factor" (FPF) is that factor by which radiation intensity is decreased as it passes through a shield. One "tenth-value thickness," that is, that thickness of the specified material that transmits a radiation dose one-tenth of that which falls upon it, provides a PF equal to 10. The protection factor concept can be applied to entire shelter structures, as well as to individual barriers or shields. For fallout shelter design, the minimum recommended PF is 40 (which require a thickness of about 1.6 times the tenth-value thickness). Either of these two measures of effectiveness may be used to describe the level of protection from fallout radiation.

Dense soils composed of discrete grains, such as sand or gravel, are said to possess the property of "dilatancy." This means that the individual soil particles normally interlock and that if a shear stress is applied to the soil, then the soil must expand slightly if the grains are to ride over each other and move. Therefore, when these soils are subjected to pressure, they develop significant shear strength. If a relatively flexible container is buried sufficiently deep in such a soil and the soil is subjected to pressure such as that from blast overpressure, the container will deflect slightly (yield) and the load will be partially transferred to the soil. The soil is said to "arch" and to partially carry the load around the container. The container can be a blast shelter constructed of corrugated metal pipe, wood, fiberglass, or even reinforced concrete. Buried concrete box structures also exhibit this type of dynamic soil–structure interaction [3].

The ability of buried timber shelters and buried corrugated metal culverts to survive very high overpressures is due to the phenomenon of earth arching. When these types of structures are buried in the right type of soil, the applied blast load is partly carried by the soil. Understanding of this phenomenon was improved in the 1960s and 1970s. Although models useful for two-dimensional calculations have been developed, they are not useful for predicting failure pressures, except to recognize that these pressures are large.

Design techniques are covered in a variety of manuals, all of which will produce shelters with a very high confidence of effectiveness. However, the reliability of design is usually attained at the cost of great conservatism and excessive expense. Significant savings on the cost of blast-resistant structures can be achieved by making use of the most advanced design techniques, such as yield-line theory or dynamic finite element analysis software (or FLAC software, etc.), and by making maximum use of improved understanding of dynamic soil–structure interactions such as earth arching [3].

Shelter from the effects of nuclear weapons may be considered quite a mature technology. Shelters can be designed from a variety of materials at a variety of costs and can be expected to function reliably with high confidence. The central problem of shelter construction programs for the United States is that any cost per occupant must be multiplied by approximately 240 million occupants (160 million with blast protection and 80 million with fallout protection). Even the best possible designs will entail significant costs for permanent blast shelters. We know how to build shelter, but we have not solved the political problem of allocating the finances to get it built.

A shelter is but one link in a whole chain of measures necessary to enable a society to survive and recover from a catastrophe, particularly a nuclear war. The spectacular pyrotechnics of nuclear weapons make it easy to forget that preparations other than shelter are required for the ultimate survival of society. The problems of water supply and food supply for survivors are only the most immediate of these problems. Equally important in the long run is the reestablishment of food production and production of the vital necessities, such as clothing, shelter, and transportation. These needs will, in turn, require establishment of some type of economy and government authority.

The threat is changing from large-yield weapons fired against urban areas to intermediate-range weapons targeted with greatly improved accuracy on more specific industrial concentrations and military targets. The advent of the cruise missile means that any active defense system is apt to cause many nuclear explosions outside of currently recognized risk areas. The biological weapon threat may also to be reconsidered. Finally, the upsurge in terrorism holds the potential for large-scale threats against civilian populations involving nuclear, biological, or chemical weapons. Let us not forget a basic principle of crisis management, do not be panicky and do not spread panic.

12.3 SHELTER OPTIONS

Table 12.1 briefly summarizes the possible different approaches to the production of shelters, their approximate costs, and their major advantages and disadvantages. Although the different shelter techniques are presented as options, any real program obviously would include a mixture of shelter techniques, depending on the threat and conditions in the area under consideration.

For completeness, a "zero option" is included in which nothing is spent for shelter space. Under these circumstances, people would make the best use of an available basement or other indoor space in a crisis. With no expenditure even for education or marking of shelters, use of the existing space would be relatively inefficient and the casualties in an all-out attack would be very high. Of more interest to the political component of our society, the nation would be much more subject to nuclear blackmail under these circumstances.

Option 11 (Table 12.1), however, is the construction of interconnected tunnel shelters under cities at considerable depth of overburden. This has apparently been carried out to a considerable extent by the Chinese under their major cities but at a shallow depth of overburden [2]. The cost is high: approximately $2000 per space; if the program were poorly managed, it could escalate to several times that amount. The system could be used only in areas of very high-density population because the cost per space would be prohibitive in areas of low-population densities.

The program produces the Rolls-Royce of civil defense shelters. Given enough entryways, people can enter the system in a few minutes. The interconnectedness of the system provides excellent protection against rubble and fire; if one entryway or ventilation air intake is blocked or in a fire, it can be simply closed off and air drawn through the tunnel from other unblocked ventilation intakes. The system also permits movement within the shelter system so that a very high-density population can be moved out to lower density areas through very long bypass tunnels.

TABLE 12.1 Shelter Options in the United States [2]

No.	Option	Cost per Space ($ US)	No. of Potential Spaces (Million)	Advantages	Disadvantages
0	Do nothing	0	—	No cost	Vulnerable to nuclear blackmail; very high casualties
1	Best available shelter	1	240	Very low cost	High casualties
2	Crisis upgrading	1–20	240	Very low cost before crisis; low cost during crisis	Requires 1 week warning; use of private property; possible clean-up cost; some evacuation required
3	Expedient shelter	1–20	240	Very low cost for planning; low cost during crisis; good protection	One week warning required; some evacuation required; short life of shelter
4	Fallout shelter in new construction	0–20	240	Low cost	No help for risk area; requires legislation; may require evacuation; long deployment time
5	Mines (modify quarrying near cities)	10–100	40–100	Moderate cost; good protection	Not applicable to all cities; 2 to 15 years deployment time
6	Earth-sheltered structure	60–300	160	Moderate cost; 2 years deployment time	Requires legislation, home sharing, blast upgrading
7	Dual-use basement in new construction	250–750	240	Low cost	Rubble and fire in central cities; 5 to 10 years deployment time; requires some evacuation

(Continued)

TABLE 12.1 Shelter Options in the United States [2]—cont'd

No.	Option	Cost per Space ($ US)	No. of Potential Spaces (Million)	Advantages	Disadvantages
8	Swiss basement shelter in new construction	350–500	240	Good protection; little warning required	Long deployment time; rubble and fire problem
9	Retrofit family shelter	500–2500	240	Two years deployment time; good protection; little warning required	High cost; not applicable in central cities
10	Retrofit dedicated blast shelters (30–50 psi)	1500–2500	200	Good protection; 2 to 5 years deployment time	Very high cost; land requirements in some areas
11	Tunnel shelters under cities	2000–5000	100	Good protection; little warning required; reduced rubble and fire problem; maximum population density	Very high cost; long deployment time

12.4 DESIGN OF SHELTERS

12.4.1 Basement Shelters

Basement spaces also offer some protection against blast. This protection has been analyzed extensively over the years by Longinow [4]. Figure 12.1 shows the indicated survival probabilities as a function of blast overpressure for different types of concrete basement construction in large buildings. There is general agreement that the principal cause of casualties in these types of buildings is collapse of the basement ceiling under blast loading. High-pressure air jetting into unprotected basement entrances can produce significant casualties among people who are standing, but relatively few among those who are lying down and out of the direct line of the entryway [4–7]. At the low blast pressure at which nonupgraded basements will collapse, initial nuclear radiation is not a significant problem.

There are some exceptions to the effectiveness of the basement shelter as an in-place shelter option. In highly built-up central business districts, where buildings of four or more stories may be constructed side by side, blast waves of only a few psi pressure will produce rubble many meters deep. Longinow and colleagues [8] estimated that fire in this debris might occur and could endanger survivors in basement shelters covered by rubble.

The other problem with low-overpressure shelters in risk areas is that of overlapping blast patterns from multiple weapons. An unsheltered person has only about a 50% chance of survival at the 0.3-atm (4-psi) overpressure range. In an area attacked by a single weapon, there is a significant improvement in the likelihood of survival for a person inside a 0.7-atm (10-psi) shelter: the mean lethal overpressure increases from about 0.3 to 0.7 atm (4–10 psi), reducing the lethal radius of the weapon by a factor of 2 and the lethal area of the weapon by a factor of 4.

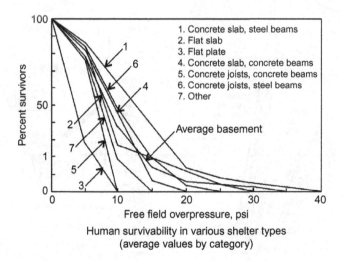

1. Concrete slab, steel beams
2. Flat slab
3. Flat plate
4. Concrete slab, concrete beams
5. Concrete joists, concrete beams
6. Concrete joists, steel beams
7. Other

Average basement

Human survivability in various shelter types
(average values by category)

FIGURE 12.1 Survival probability in basements [2].

There is an equal reduction in the probability of not surviving. However, if a pattern of weapons is laid down on an area, it will be designed to subject substantially the entire area to the overpressure—selected by the designer of the attack to produce the desired effect. In such a case, very little of the target area is subjected to over-pressures between 0.3 and 0.7 atm (3 and 10 psi or 0.03 and 0.07 MPa), where these best available shelters can improve survivability.

Single-weapon attacks are conceivable from terrorists, private armies, third-world countries with (emotionally) mentally unstable leaders, or accidental or unauthorized launches from other nuclear powers. Against this type of contingency, a factor of 4 reduction in casualties is well worth attaining. The existing system of best available shelter would provide very few additional survivors in the event of an all-out attack on population centers at present.

12.4.2 Expedient Shelter Designs

The simple expedient shelter against a single-weapon attack is a foxhole. A small pit is dug [75 cm (30 in.) wide and 120 cm (4 ft) deep], which provides the occupant with a fallout protection factor of approximately 40. If covered with a canopy, for example, a shower curtain or bed sheet, that is kept free of fallout, the occupant sitting in the foxhole could have a protection factor of about 200. Many expedient shelter designs have been developed. Concepts for covered trenches made with doors or planks were published in the early 1960s. Figure 12.2 illustrates some different techniques used in constructing

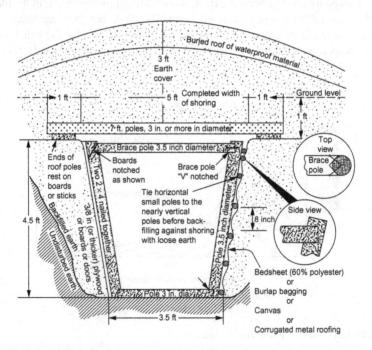

FIGURE 12.2   Methods for shoring a trench shelter [2].

a shored trench shelter. This type of construction has withstood blast pressures of 1.4 atm (20 psi) and provides a fallout protection factor (given sufficient attention to entrances) of about 200.

Modern tunnel or cavern type shelters are discussed in the form of case histories in the following section.

12.5 CASE HISTORIES

12.5.1 Civil Defense Shelter in Singapore

A civil defense bomb shelter has been planned with the warehouse caverns in the Bukit Timah granites in Singapore. The scheme consists of five warehouse caverns and an optional additional cavern as a civil defense bomb shelter [9].

The layout of the underground shelter is shown in Figure 12.3. The main shelter cavern is 15 m wide, 8 m high, and 80 m long. Separate side rooms are provided to accommodate toilets, hospital facilities, and technical installations. The defense shelter has two entrances: one from the warehouse access tunnel and the other through a separate inclined shaft. Both entrances are equipped with blast-proof doors and gas sluices.

The 9600-m^3 excavated volume of the shelter facility provides 4.8 m^3 per person for a capacity of 2000 persons. This is adequate according to current Norwegian civil defense regulations.

During peacetime, the civil defense shelter can be used for storage purposes. Alternatively, it might be converted for recreational purposes, such as a shooting range or squash, badminton, or tennis courts.

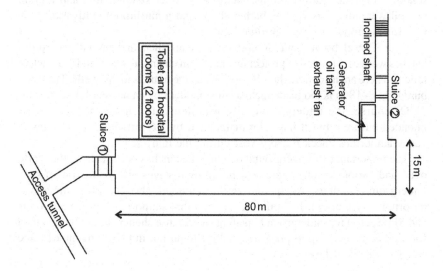

FIGURE 12.3 Layout of proposed civil defense shelter [9].

12.5.2 Beijing's Underground City

Beijing's underground city, a bomb shelter just beneath the ancient capital's downtown area, has been virtually forgotten by local citizens for more than 20 years, despite being well known among foreigners since it officially opened in 2000 (http://english.hanban.edu.cn/english/travel/125961.htm).

The underground city has also been called the Underground Great Wall, as it had the same purpose: military defense. This complex is a relic of the Sino–Soviet border conflict in 1969 over Zhenbao Island in northeast China's Heilongjiang River, a time when chairman Mao Zedong ordered the construction of subterranean bomb shelters in case of nuclear attack.

The tunnels, built from 1969 to 1979 by more than 300,000 local citizens and even schoolchildren, extend for over 30 km and cover an area of 85 km^2, 8 to 18 m under the surface. It includes around a thousand anti-air raid structures.

Centuries-old city walls and towers that once circled ancient Beijing were destroyed to supply construction materials for the complex. The old city gates of Xizhimen, Fuchengmen, Chongwenmen, and others remain in name only—only two embrasure watchtowers from Zhengyangmen and Deshengmen survived.

In the event of attack, the plan was to house 40% of the capital's population underground and for the remainder to move to neighboring hills; it is said that every residence once had a secret trapdoor nearby leading to the tunnels.

There is no authoritative information on how far the mostly hand-dug tunnels stretch, but they supposedly link all areas of central Beijing, from Xidan and Xuanwumen to Qianmen and Chongwen districts, to as far as the Western Hills.

They were equipped with facilities such as stores, restaurants, clinics, schools, theaters, reading rooms, factories, a roller skating rink, and a grain and oil warehouse, as well as barber shops and a mushroom cultivation farm for growing foods that require little light.

Over 2300 elaborate ventilation shafts were installed, and gas and waterproof hatches were constructed to protect insiders from chemical attack and radioactive fallout. There are also more than 70 sites inside the tunnels to dig wells. The overburden of 8–18 m is too inadequate against modern mega nuclear bomb attacks.

Of course, the underground city was thankfully never needed for its intended purpose, but it has been maintained by city officials. Water conservancy authorities check it every year during the rainy season.

The importance of underground defense shelters has been realized the world over, and various countries are now in different stages of construction of such underground shelters. Adequate overburden depth above nuclear shelters is unfortunately ignored. It should be ensured that support systems (resin bolts and SFRS, etc.) for underground openings as defense shelter should be designed for high dynamic support pressures and it should fail in plastic manner for the safety of people below.

REFERENCES

[1] http://www.espionageinfo.com/Ul-Vo/ - Underground Facilities, Geologic and Structural Considerations in the Construction. Accessed in 2009.

[2] Chester CV, Zimmerman GP. Civil defense shelters: a state-of-the-art assessment. Tunnel Undergr Space Technol 1987;2(4):401–28.

[3] Garg SK. Non-linear transient dynamic soil-structure interaction with special reference to blast loading. Ph.D. thesis, IIT Roorkee, India; 1998. p. 324.

[4] Longinow A. Survivability in a nuclear weapon environment (final report). Defense Civil Preparedness Agency, IITRI project J6427 (AD-A076026). Chicago, IL: IIT Research Institute; 1979 [in Chester CV, Zimmerman GP (1987)].

[5] Longinow A. Probability of people survival in a nuclear weapon blast environment (final report). FEMA, IITRI project J6479-FRE (AD-A087600). Chicago, IL: IIT Research Institute; 1980 [in Chester CV, Zimmerman GP (1987)].

[6] Longinow A, Hahn E, Wiedermann A, Citki S. Casualties produced by impact and related topics of people survivability in a direct effects environment (final report). Defense Civil Preparedness Agency, IITRI project J6303 (AD-A011108). Chicago, IL: IIT Research Institute; 1974 [in Chester CV, Zimmerman GP (1987)].

[7] Longinow A, Wiedermann A, Citko S, Iwankiw N. Debris motion and injury relationships in all hazard environments (final report). Defense Civil Preparedness Agency, IITRI project J6334 (ADA030815). Chicago, IL: IIT Research Institute; 1976 [in Chester CV, Zimmerman GP (1987)].

[8] Longinow A, Waterman TE, Takata AN. Assessment of combined effects of blast and fire on personnel survivability. IITRI project J6483 (AD-A117958). Chicago, IL: IIT Research Institute; 1982 [in Chester CV, Zimmerman GP (1987)].

[9] Wallace JC, Ho CE, Bergh-Chrlatensen J, Zhao J, Zhou YX, Choa V. A proposed warehouse-shelter cavern scheme in Singapore granite. Tunnel Undergr Space Technol 1995;10(2):163–7.

Chapter 13

Underground Storage of Ammunitions and Explosives

The art of progress is to preserve order amid change and preserve change amid order.

A.N. Whitehead

13.1 GENERAL

Underground storage of ammunition and explosives is gaining momentum in view of increased terrorist and sabotage activities. Moreover, underground storage helps in releasing surface land, which can be used for other purposes more efficiently. Underground ammunition storage sites should be located in sound rock. A storage site may consist of one or more storage chambers with usually one access tunnel in each chamber. The number of chambers depends on prevailing topographical and geological circumstances and safety aspects in the environment of the storage site. Potential blockage should be considered for multichamber sites.

The thickness of the rock formation surrounding an underground storage site should be designed so that cratering hazards, in case of an explosion, may be practically excluded. In such a situation the only significant external hazards will be the ground shock and the explosion (air pressure) effects coming from the adit tunnel. Effects coming from the adit may be reduced considerably by means of structural measures in or in front of the tunnel or even eliminated by the installation of tunnel-closing devices. Adequate separations and tunnel-closing devices shall be used to prevent the propagation of an explosion from chamber to chamber. Provided the access openings are adequately strong, underground storage is relatively well protected against enemy attack.

Geological aspects have a great influence on building costs and, in terms of the construction cost alone, underground storage is often more costly than aboveground storage. However, when estate, operating, maintenance, and lifetime costs are considered, at least for larger underground facilities, they may be less than for comparable size aboveground facilities. Generally, chambers measuring from 100 to 200 m in length with a volume between 5000 and 15,000 m^3 are the most economical. This provides a total gross capacity between 1000 and 2000 tonnes of ammunition. The length of the access tunnel may be 50 to 150 m, depending on topographical conditions and the desired rock thickness (cover).

A report of the North Atlantic Treaty Organization (NATO) published in May 2006 is the main reference material for preparing this chapter.

13.2 EXPLOSION EFFECTS IN UNDERGROUND AMMUNITION STORAGE SITES

The blast wave originating from an explosion in an underground storage chamber will surge through the rock formation as ground shock and will escape as blast through the access tunnel into the open air. The strong confining effect of an underground storage site and the large amount of hot explosion gases generated will produce a relatively constant high pressure in the chamber. This pressure may break up the rock formation and produce a crater. The kinetic energy (dynamic pressure impulse) of the blast in the main passageway is very high compared to an explosion in free air. Objects such as unexploded ordnance, rock, gravel, equipment, and vehicles will be picked up and accelerated up to velocities of several hundred meters per second before leaving through adits. In addition, engineered features can collapse and cause debris hazards. Break-up of the cover will cause projection of a heavy ejection of rock and earth in all directions on the surrounding surface area.

The explosion gases will surge at a high velocity through the access tunnel into the open air where they will burn completely. The escaping gases will carry along ammunition, rock debris, installations, and lining onto surrounding areas.

A disturbance near the surface of the ground will emit compression P-waves, shear S-waves, and Raleigh surface R-waves in a semi-infinite elastic medium. Deeply buried disturbances will emit only P-waves and S-waves, but in the far field, interface effects will result in R-waves on the surface. For all of these wave types, the time interval between wave front arrivals becomes greater and the amplitude of the oscillations becomes smaller with increasing standoff distance from the source.

The first wave to arrive is the P-wave, the second the S-wave, and the third the R-wave. P-waves and S-waves cause minor tremors, as these waves are followed by much larger oscillations when the R-wave arrives. The R-wave is the cause of the major tremor because (i) about two-thirds of the ground shock energy at the source goes into the R-wave and (ii) the R-wave dissipates much less rapidly with distance than either the less energetic P-wave or S-wave. P-waves and S-waves dissipate with distance r to a power of r^{-1} to r^{-2}. At the surface, P-waves and S-waves dissipate with distance as r^{-2}, while R-waves dissipate with distance as $r^{-0.5}$. The greater energies being transmitted by R-waves and the slower geometric dissipation of this energy cause R-waves to be the major tremor (the disturbance of primary importance for all disturbances on the surface).

Small-Scale Model Tests and Validity of Scaling Laws

A portion of the blast energy from an underground detonation is used to compress surrounding geological media. This allocation of energy should be considered when evaluating the experimental results of underground tests.

Small-scale model tests constructed of nonresponding materials do not exhibit nonlinear energy loss effects typical of an underground explosion. Therefore, air blast results from nonresponding models tend to be conservative for predicting hazards that would occur in an actual underground event. Despite this limitation, small-scale model tests are still of value for design purposes.

Zhao and colleagues carried out experiments to study the properties of Bukit Timah granite under dynamic conditions [1]. Results of the studies showed that rate (time) and stress wave effects are two important factors influencing rock dynamics problems in addition to all factors influencing traditional rock mechanics problems.

It is believed that numerical modeling with the proper understanding of the basic dynamic properties of the rock materials, rock joints, and rock mass is the key approach in studying the shock wave propagation in a fractured rock mass, the response of the rock mass, and the stability of rock structures under dynamic loads. Small-scale field experiments and laboratory tests provide the necessary input parameters for numerical modeling. Limited large-scale field tests shall be conducted mainly to calibrate the numerical simulation.

13.3 ADVANTAGES AND DISADVANTAGES OF UNDERGROUND STORAGE

Advantages of underground storage are:

- A smaller total land area is required than for aboveground storage.
- A high degree of protection is afforded against bombing or terrorist attack.
- The area is easier to camouflage and to guard than an aboveground area.
- In case of an incident in an underground chamber, damage to ammunition in other chambers is preventable. Damage to ammunition in aboveground buildings, other than earth-covered magazines, is usually more extensive.
- The temperature in underground storage sites is almost constant. The deleterious aging effects on ammunitions and explosives caused by extreme temperatures and temperature cycling are mitigated.
- Effects of sand, snow, and ice, which may cause difficulties in aboveground storage, may be avoided.
- Inherent protection may be achieved against external fire.
- Estate costs, as well as maintenance and operation, may be less costly as for an aboveground storage site, thus more than offsetting the construction costs.

- Rock caverns may have an almost unlimited life span compared to above-ground structures, which are subject to the effects of weather.

Disadvantages, however, of underground storage may be:

- The choice of localities is restricted.
- Costs of original excavation or modification of an existing excavation and the installation and maintenance of special equipment may increase the initial costs of underground storage over that of aboveground storage.
- Extra handling equipment may be required.

Work Prohibited in Underground Storage Sites

The opening of packages or the removal of components from unpacked ammunition or similar operations should be prohibited in the storage chamber, but could be done in the loading/unloading dock or in a separate chamber if suitable measures are taken to prevent a propagation of a gas explosion into the storage chambers.

13.4 STORAGE LIMITATIONS

Limitations on underground storage are as follow.

13.4.1 Ammunition Containing Flammable Liquids or Gels

Ammunition containing flammable liquids is only permitted in underground storage sites if proper protection against fuel leakage is established. Possible energy release of a stochiometric combustion should be considered as part of the total energy release. Stochiometric or theoretical combustion is the ideal combustion process during which a fuel is burned completely. Multichamber sites should be arranged and/or sealed in such a way that a fuel-fire or gas explosion should not increase the likelihood of reaction in neighboring chambers more than established through interior distances to prevent detonation transfer.

13.4.2 Ammunition Containing Toxic Agents

Ammunition containing toxic agents should only be stored under special provisions because of the difficulties of decontamination underground.

13.4.3 Suspect Ammunition and Explosives

Suspect ammunition and explosives should not be stored.

13.4.4 Ammunition Containing Pyrotechnics

Ammunition containing pyrotechnics, such as illuminating, smoke, and signal ammunition, could, in some cases, be more vulnerable to mishaps or self-ignition, thereby increasing the likelihood of an accident. The decision to store ammunition that contains pyrotechnics underground must be made on a site-specific basis and provisions must be taken to mitigate the peculiar hazards of pyrotechnic materials.

13.4.5 Ammunition Containing Depleted Uranium

The slight radioactivity and chemical toxicity that would result from an accidental fire or explosion should be assessed and accepted before ammunition containing depleted uranium is permitted in underground sites.

13.5 DESIGN REQUIREMENTS OF UNDERGROUND AMMUNITION STORAGE FACILITY

Planning of underground storage facilities must account for site conditions, storage requirements, and operational needs. Only when these are established can the design be developed. An optimal compromise between the sometimes contradictory demands for planning, construction, and operation of storage sites must consider safety, military, and cost requirements.

13.5.1 Safety Requirements

Operational procedures should be planned and conducted so that explosives mishaps are prevented to the best extent possible. Facility configurations are to be designed so that if an explosive mishap should occur, its hazards are mitigated to acceptable levels. Safety efforts essential for ammunition storage sites include:

- Surveillance and maintenance to ensure that only safe ammunition is stored.
- Well-designed and environmentally controlled chambers and facilities to protect ammunition against unintended events.
- Suitable structural designs and operating procedures (e.g., doors and guards) to protect ammunition against deliberate action by third parties
- Structural designs and operating procedures to
  - Mitigate explosion propagation outside the area of initial occurrence
  - Provide desired levels of personnel, facility, and asset protection
- Construction and operation of ammunition storage sites should only be entrusted to qualified and trained personnel who have clearly defined responsibilities.
- Evaluate a suitable location for installation taking into account the site-specific use of surrounding structures (inhabited buildings, roads, etc.).

13.5.2 Military Requirements

Functional requirements that dictate the geographical location of a storage site or its storage and transfer capacity may sometimes run counter to desirable safety considerations, thereby requiring innovative designs to provide required levels of explosive safety. Military requirements often involve protection against enemy weapons, intruder protection, etc.

13.5.3 Financial Aspects

The lifetime cost of a storage facility (construction, operation, and maintenance) should be considered during the planning phase. Where possible, designs should be selected that minimize total cost while providing required safety and operational capabilities.

13.5.4 Humidity Control and Ventilation

High humidity may be a problem in underground sites. Dehumidifying equipment may then be necessary to control relative humidity to about 60%. Chambers may be lined with concrete or coated fabric for better control of humidity. The roof lining should be strong enough to withstand minor rock falls.

The type of transportation equipment (used) may govern ventilation requirements. Ventilation shafts to the exterior should be designed to prevent trespass and sabotage.

13.5.5 Electric Installations and Equipment

- Electric installations and equipment for underground storage sites should conform to national standards of the host nation.
- An emergency lighting system should be installed. Transportable battery-operated lights of an appropriate standard should be provided and kept at suitable points.
- A portion of the personnel employed underground should be equipped with hand lamps of an appropriate standard.

13.5.6 Lightning Protection

An underground storage site does not normally require a system of protection against lightning. Metal and structural parts of the site that have less than 0.6 m cover should be protected as for an aboveground site. However, each underground storage site should be considered individually to take account of possible conducting faults in the cover.

13.5.7 Transport and Handling Equipment

Rail vehicles, road vehicles, mobile lifting or stacking appliances, and cranes of the fixed or gantry type, when operated electrically or by diesel engine, may be permitted in underground storage sites subject to the following conditions.

- Electrical equipment should conform to national standards of the host nation for underground storage sites.
- Diesel-operated equipment should be fitted with an effective means of preventing sparks or flames from exhaust outlets. Any portion of the exhaust system or exposed parts of the engine, which may develop a surface temperature exceeding 100°C, should be suitably screened to ensure that all exposed surfaces are below that temperature. If the engine is to be kept running during loading and unloading within the storage site, it should conform to the host nation standards for underground (confined space) operations.
- The flash point of the fuel oil for diesel engines should not be less than 55°C. Fuel tanks should be filled only at authorized places and no spare fuel should be carried.
- Where fuel oil-filling stations are authorized in the underground area, fuel should be taken underground in strong closed containers in quantities not exceeding that required for one working day. The filling station should have a concrete floor with a sill of sufficient height to contain the quantity of fuel authorized to be stored there.

13.5.8 Fire-Fighting Equipment

Equipment should conform to national standards of the host country with particular consideration given to the following.

- Reduce the probability that a small fire will escalate by installing an automatic smoke-detecting and fire-extinguishing system.
- Consideration should be given to protecting reserve water tanks from potential explosive effects.
- An alarm system should be provided to operate throughout the whole area, both above and below ground.
- In air-conditioned sites or in sites provided with forced ventilation, the need to shut these down upon outbreak of fire must be considered.
- Fire-fighting equipment retained underground should be positioned for accessibility and potential use.
- For large underground areas, detector devices, to specify the location of a fire, and communication capabilities, to issue instruction throughout the underground facility, should be installed.
- Self-contained breathing apparatus and training in its use are essential for underground fire-fighting or rescue operations, etc.

13.6 FACILITY LAYOUT

13.6.1 Design Guidelines (Source: www.dir.ca.gov/Title8/5258.html)

- Explosive materials stored underground shall be located so that if they detonate or burn, the escape route for the employees will not be obstructed.

- Magazines shall be at least 200 feet (60 m) from active underground workings, 50 feet (15 m) from other magazines, and at least 25 feet (7.5 m) of solid ground separation from any haulage way used for any purpose other than the transportation of explosive materials. Any timbers within 25 feet (7.5 m) of any magazine shall be made fire resistant.
- Explosive materials shall not be stored in an underground work area during tunneling and construction operations.
- In magazines where explosive materials may become damp, electricity may be installed for drying purposes. Electrical equipment used shall comply with class II, division I, hazardous locations, electrical safety orders. Electrical wiring shall be kept at least 5 feet (1.5 m) from explosive materials. No other electrical wiring shall be permitted within 5 feet (1.5 m) of any magazine.
- Underground storage magazines shall be conspicuously marked with the word "EXPLOSIVES" in red letters at least 4 inches (10 cm) high and with a 5/8-inch (1.6 cm) stroke on a white background.
- Combustible rubbish shall not be permitted within 100 feet (30 m) of any underground magazine.
- Detonator storage magazines shall be of the same construction as explosive storage magazines and shall be separated by at least 50 feet (15 m) from other magazines.

13.6.2 Underground Chambers

- A single-chamber facility with a straight access tunnel leading from the chamber to the portal is a "shotgun" magazine because blast and debris behave as if fired from a gun. More complex facility layouts will provide reductions in exit pressures.
- The side on pressure, side on pressure impulse, dynamic pressure, and dynamic pressure impulse decrease as the volume increases.
- Distributing ammunitions over several storage chambers may control the size of an initial explosion. Proper separation or hazard-mitigating construction can limit subsequent damage.

Underground storage facilities may consist of a single chamber or a series of connected chambers. The chamber(s) may be either excavated or natural geological cavities. Figures 13.1 and 13.2 illustrate general concepts for several possible configurations of underground facilities.

13.6.3 Exits

Exits from underground storage sites should not emerge where they direct blast, flame, and debris hazards to exposed sites, such as other entrances, buildings, or traffic routes.

Connected chambers and cave storage sites should have at least two exits. Exits should be separated by at least the chamber interval.

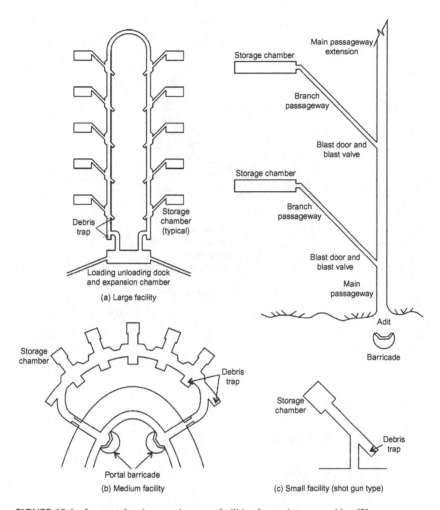

FIGURE 13.1   Layout of underground storage facilities for storing ammunition [2].

13.6.4 Branch Passageways

When a main passageway has one exit, branch passageways should be inclined at an angle where they join the main passageway to direct the flow field toward the exit. This inclination should provide for vehicle access. Angles between 40 and 70° are normally appropriate.

The rock thickness between the chamber and the main passageway should be at least equal to or greater than the chamber interval, as otherwise an explosion in a chamber might destroy the main passageway and prevent access to stocks of ammunition and explosives in the other chambers.

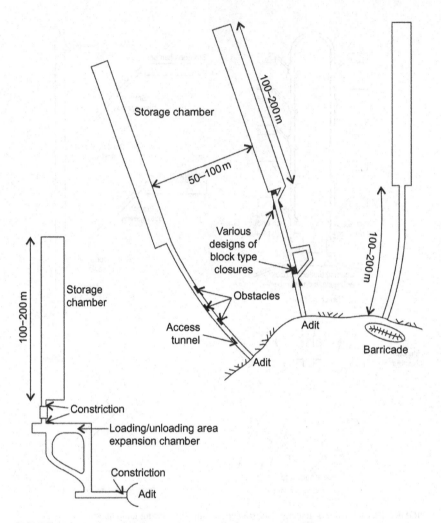

FIGURE 13.2   Layout of underground storage site for ammunition [2].

13.6.5 Blast Closures

High-pressure closures are large blocks made of concrete or other materials that can obstruct or greatly reduce the flow of blast effects and debris from an explosion or into a storage chamber. For chamber-loading densities of about 10 kg/m³ or above, closure blocks will contain 40% or more of the explosion debris within the detonation chamber, provided the block is designed to remain intact. If a closure block fails under the blast load, it will produce a volume of debris in addition to that from the chamber itself. However, because the mass and inertia of the block are sufficient to reduce the velocity of the primary debris greatly, the effectiveness of other debris-mitigating features, such as debris traps, expansion chambers, and barricades, is increased. Debris traps and expansion chambers intended to entrap debris must be designed to contain the ejection of the full potential volume of debris, based on possible explosion of the maximum capacity of the largest storage chamber.

These debris mitigation features were investigated by tests described by Davis and Song [3]. These tests showed that such measures can be very effective; however, no quantitative figures for a reduction of the adit debris throw were derived. Furthermore, it was shown that a proper design of mitigation measures is very important. Sample drawings of features that proved to be effective for the tested configurations are proposed by the U.S. Army Corps of Engineers [4].

An alternative, full-scale tested design for a high-pressure closure device, the Swiss-Klotz design, is shown in Figure 13.3. This device is highly effective up to chamber-loading densities of $28 \, \text{kg/m}^3$. A special advantage of this design is that it is movable and can be closed during times when access to the storage chamber is unnecessary.

In case of an explosion inside the storage chamber and a Klotz in a closed position, practically all of the hazardous debris, as well as the explosion gases, will be trapped inside the storage chamber, thereby reducing these hazardous effects to virtually insignificant levels. In case the Klotz is in an open position, it will be pushed into the closed position by the explosion gases within approximately 100 ms, letting only a small fraction of the total amount of debris and gases pass.

In any case, using a properly designed high-pressure closure device in conjunction with a portal barricade will lower the debris hazard to a level where specific debris quantity–distance throw considerations will not be required. Other combinations of mitigation features will also reduce adit debris throw to a great extent. The remaining adit debris hazard has to be assessed based on the actual facility layout and quantified by means of suitable tests.

Blast doors that are protected from primary fragments have proven effective for loading densities up to $10 \, \text{kg/m}^3$.

13.6.6 Expansion Chambers

Expansion chambers are so named because of the volume they provide for the expansion of detonation gases behind the shock front as it enters the chamber from a connecting tunnel. Some additional degradation of the peak pressure at the shock front occurs as the front expands into the chamber and reflects from the walls.

Expansion chambers have other practical purposes. They serve as loading/unloading chambers, as weather-protected areas for the transfer of ammunitions from trucks to storage chambers, and as turnaround areas for transport vehicles.

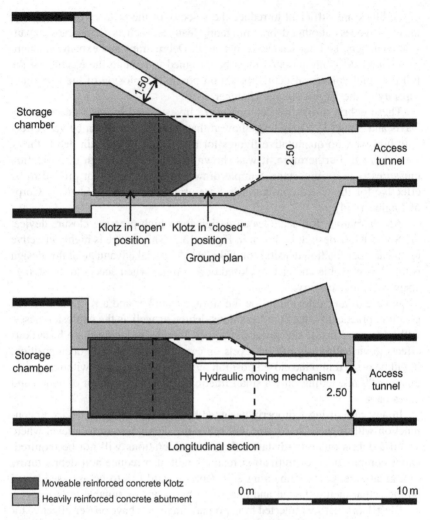

FIGURE 13.3 High-pressure closure device in Swiss-Klotz layout [2].

Figures 13.4 and 13.5 illustrate underground facilities with and without expansion chambers.

13.6.7 Constrictions

Constrictions, which may be used for mitigating explosives hazards, are short lengths of tunnel with a reduced cross-sectional area.

A constriction at a chamber entrance reduces the magnitude of air blast and thermal effects entering chambers near one in which an explosion might occur.

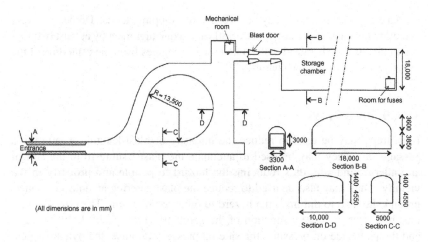

FIGURE 13.4 Typical magazine with 2300-m^3 expansion chamber as blast trap [2].

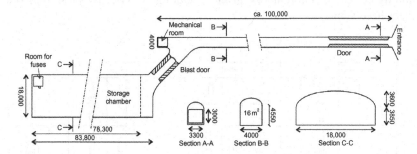

FIGURE 13.5 Typical magazine with straight tunnel and no expansion chamber [2].

A constricted chamber entrance also reduces the area and hence the size of a blast door installed to protect the chamber contents.

A constriction intended to reduce air blast issuing from an exit of an underground storage facility should be located within five tunnel diameters of the exit.

Although constrictions located more than five tunnel diameters from exits will reduce pressures by delaying the release of energy [5,6], their effects on pressure versus distance must be considered on a site-specific basis.

13.6.8 Debris Traps within Underground Facility

Debris traps are excavations in the rock at or beyond the end of sections of tunnel designed to catch debris from a storage chamber detonation. Debris traps should be at least 20% wider and 10% taller than the branch passageway from the chamber whose debris it is intended to trap, with a depth (measured along the shortest wall) of at least one tunnel diameter.

An expansion chamber may be effective for trapping debris. Tunnels entering or exiting the chambers must either be offset in axial alignment by at least two tunnel widths or its axis must be offset at least 45° degrees from the centerline of the tunnel associated with the chamber [4].

13.6.9 Blast Traps

Blast traps may be used to reduce the intensity of blast leaving or entering a passageway. They may be used to attenuate the blast issuing from the adit of an underground site, thus reducing the hazard to people and property in the vicinity. They may also be used to reduce the blast entering an adjacent underground site, and to diminish the hazard to other ammunition. The effect of various blast traps will be a function of the geometrical design of the blast traps, and the peak side on pressure, the side on pressure impulse, the dynamic pressure, and the dynamic pressure impulse of the incident blast wave. Fixed reduction figures can therefore not be given. The design of effective blast traps is a specialized subject.

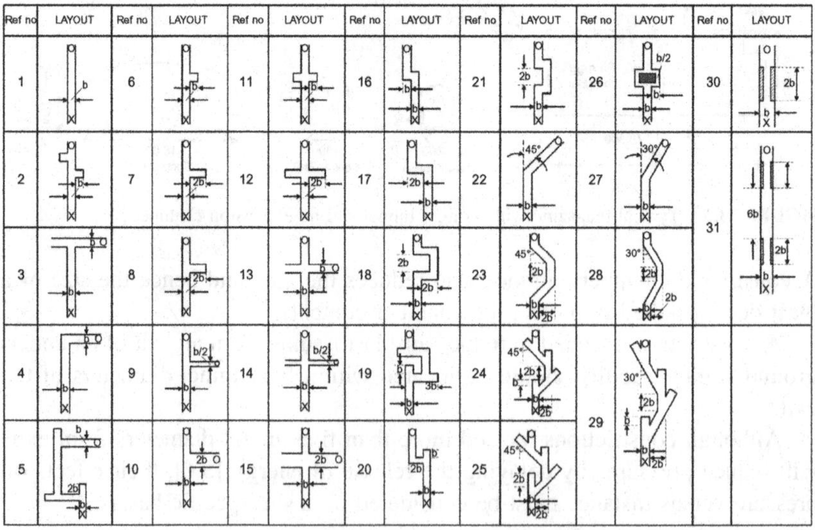

FIGURE 13.6 Various types of blast traps [2]. ((i) Turns, crossovers, obstacles, and changes of cross section can be used to reduce the peak overpressure and positive impulse of a blast in passageways. Diagrams illustrate some of the many possible designs. Blast k assumed to travel from the point indicated by a cross to that shown by a dot. Critical dimensions are indicated as multiples of passage diameter "b." (ii) Some designs have comparatively little effect, reducing the blast by only 10% compared with the straight-through passageway in Ref. No. 1, whereas others reduce the blast by as much as 80%. It is, therefore, necessary to determine the actual effect of a chosen design by measurements in a model using properly scaled and located explosive charges.)

Various types of blast traps are shown in Figure 13.6. The relative decrease of pressure and impulse, and thereby the effect of these blast traps, is, in most cases, dependent on their locations. Some of the limitations are also indicated in Figure 13.6. It is noteworthy that not all designs of blast traps are reversible.

For maximum blast reduction, the length of blast traps built as dead-end tunnels should be at least half the length of the blast wave. This may result in a considerable extension of these traps in the case of large quantities of explosives.

13.6.10 Portal Barricade

Air blast. An air blast exiting the portal of an underground facility involves directional, very intense gas flow fields along the extended centerline of the tunnel exit. Therefore, the shock wave on the extended centerline does not attenuate as rapidly as that of a surface burst. However, a barricade in front of the portal intercepts this intense flow field and directs it away from the extended centerline axis. This redirection of the flow field allows shock waves traveling beyond the portal barricade to attenuate as an aboveground-distributed source so that isobar contours become more circular. Figure 13.7 provides an example of a portal barricade.

Adit debris. A portal barricade reduces the inhabited building distance for adit debris by obstructing the path of the debris as it exits the tunnel. However, the barricade must be used in conjunction with another debris mitigating construction, such as a debris trap. The quantity–distance decisions must be made by site-specific analyses and/or testing.

To be effective, a portal barricade must be designed and located properly (Fig. 13.7).

Portal barricades for underground magazines are located immediately in front of the portal. The portal barricade should be centered on the extended axis of the tunnel that passes through the portal. For maximum effectiveness, the face of the barricade toward the portal must be vertical and concave in plan. Its central face must be oriented perpendicular to the tunnel axis, and its wing walls should be angled about 45° toward the portal. The minimum width of the central face is equal to the width of the tunnel at the portal. Wing walls must be of sufficient width so that the entire barricade length intercepts at an angle of at least 10° (to the right and left) of the extended width of the tunnel. Likewise, the height of the barricade along its entire width must be sufficient to intercept at an angle of 10° above the extended height of the tunnel [7].

Portal barricades for underground magazines must be located at a distance of not less than one and not more than three tunnel widths from the portal. The actual distance should be no greater than that required for the passage of applicable transportation equipment. This distance, as shown in Figure 13.7,

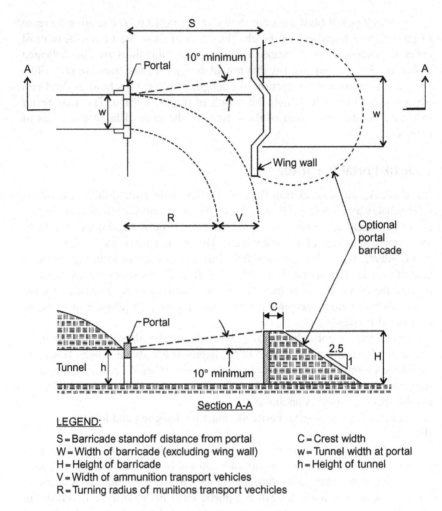

LEGEND:

S = Barricade standoff distance from portal
W = Width of barricade (excluding wing wall)
H = Height of barricade
V = Width of ammunition transport vehicles
R = Turning radius of munitions transport vechicles

C = Crest width
w = Tunnel width at portal
h = Height of tunnel

FIGURE 13.7 Portal barricade location, height, and width [2].

is based on the turning radius and operating width of the transportation equipment.

The barricade must withstand the impact of debris ejected from the tunnel and must be sufficiently robust so it does not contribute to debris hazards.

13.6.11 Interior Wall Roughness

Although the effect of wall roughness is not fully quantified, some of the differences between measured and predicted results are likely attributable to it [5,6]. Mitigating effects of wall roughness should be considered on a site-specific basis.

13.6.12 Depth of Cover above Storage Chambers

To qualify as an underground facility, the minimum distance in meters from the perimeter of a storage area to an exterior surface shall be greater than $0.10\,Q^{1/3}$, where Q is explosive weight in kilograms. (Note: This minimum distance normally, but not always, equals the thickness of the earth cover.) If this criterion cannot be met, the facility must be sited as an aboveground magazine [7].

Considerations of rock cover above storage facilities are significant for planning and evaluating underground explosive storage. It is intuitively evident that the amount of earth/rock cover over a given amount of explosive will have a significant effect in the blast and shock phenomenology in the event of detonation. In the limiting case, an explosion of almost any size that would be conceivable for conventional storage situations would produce negligible surface effects if it occurred at a depth of several kilometers. However, for practical situations, the depth of cover is a factor that must be considered quantitatively in evaluating air blast and debris effects.

The depth of cover or chamber cover thickness is the shortest distance between the ground surface and the natural rock surface at the chamber's ceiling or, in some cases, a chamber's wall. For all types of rock, the critical cover thickness required to prevent breaching of the chamber cover by a detonation (C_c) is

$$C_c = 1.0\,Q^{1/3}, \text{meters}, \qquad (13.1)$$

where C_c is in meters and explosive charge Q is in kilograms [7].

13.6.13 Chamber Separation Requirement

Minimum storage chamber separation distances are required to prevent or control the communication of explosions or fires between chambers. There are three modes by which an explosion or fire can be communicated: rock spall, propagation through cracks or fissures, and air blast or thermal effects traveling through connecting passages. Spalled rock of sufficient mass traveling at a sufficient velocity may damage or sympathetically detonate impacted ammunition and explosives in the acceptor chambers.

- *Prevention of rock spall* [7]. The chamber separation distance is the shortest distance (rock thickness) between two chambers. When an explosion occurs in a storage chamber, a shock wave is transmitted through the surrounding rock. The intensity of the shock decreases with distance. For small chamber separation distances, the shock may be strong enough to produce spalling of the rock walls of adjacent explosive storage chambers. When no specific protective construction is used, the same may be determined as follows [7]:
  - For moderate to strong rock

– For loading densities less than or equal to 50 kg/m³, the minimum chamber separation distance (D_{cd}) required to prevent hazardous spall effects is the same as Eq. (13.1), that is,

$$D_{cd} = 1.0\,Q^{1/3}, \text{meters}, \tag{13.2}$$

where D_{cd} is in meters and explosive charge Q is in kilograms. (Note: D_{cd} shall not be less than 4.6 m.)

– For loading densities greater than 50 kg/m³, the separation distance is

$$D_{cd} = 2.0\,Q^{1/3}, \text{meters}. \tag{13.3}$$

• For weak rock, at all loading densities, the separation distance is

$$D_{cd} = 1.40\,Q^{1/3}, \text{meters}. \tag{13.4}$$

• *Prevention of propagation by rock spall.* If damage to stored ammunitions in the adjacent chambers is acceptable, the chamber separation distance can be reduced to the distance required to prevent propagation of the detonation by the impact of rock spall against the munitions. For smaller distances, propagation is possible. Propagation by rock spall is practically instantaneous because time separations between donor and acceptor explosions may not be sufficient to prevent coalescence of blast waves. Unless analyses or experiments indicate otherwise, explosives quantities subject to this mode must be added to other donor explosives to determine net explosive quantity. For loading densities up to 270 kg/m³, when no protective construction is used, the separation distance, D_{cd}, to prevent explosion communication by spalled rock is

$$D_{cd} = 0.6\,Q^{1/3}. \tag{13.5}$$

When the acceptor chamber has protective construction to prevent spall and collapse (into the acceptor chamber), the separation distance must be determined on a site-specific basis but may be as low as

$$D_{cd} = 0.3\,Q^{1/3}. \tag{13.6}$$

• *Prevention of propagation through cracks and fissures.* Propagation between a donor and an acceptor chamber has been observed to occur when natural, near horizontal jointing planes; cracks; or fissures in the rock between the chambers are opened by the lifting force of the detonation pressure. Prior to construction of a multichamber magazine, a careful site investigation must be made to ensure that such joints or fissures do not extend from one chamber location to an adjacent one. Should such defects be encountered during facility excavation, a reevaluation of the intended siting is required [7].

• *Prevention of propagation through passageways.* Flame and hot gas may provide a delayed mode of propagation [7]. Time separations between events in the donor chamber and the acceptor chamber by this mode will be sufficient to prevent coalescence of blast waves. Consequently, siting is based on each chamber's net explosive weight for quantity distance. To protect assets, blast and fire-resistant doors may be installed within multichambered facilities. Evaluations for required chamber separations due to this propagation mode should be made on a site-specific basis.

- Chamber entrances at the ground surface, or entrances to branch tunnels off the same side of a main passageway, shall be separated by at least 15 feet (4.6 m).
- Entrances to branch tunnels of opposite sides of a main passageway shall be separated by at least twice the width of the main passageway.

• *Propagation by flame and hot gas through cracks and fissures.* Consideration must be given to the long-duration action of the explosion gas. These quasistatic forces may form cracks in the rock that extend from the donor to an adjacent (acceptor) chamber, thus making it possible for hot gases to flow into this chamber and initiate an event. Significant factors for this mode of propagation include the strength of rock, the existence of cracks formed before the explosion incident, the type of barriers in cavern storage sites, the cover, and the loading density in the chamber. This mode of propagation must be considered when final decisions about chamber separation distances are made.

Thus, because of these cracks and fissures, propagation may occur beyond

$$D_{cd} = 0.3\,Q^{1/3}. \tag{13.7}$$

Equation (13.7) is similar to Equation (13.6).

13.7 SYMPATHETIC DETONATION BY ROCK SPALL

The minimum separation distance to prevent a sympathetic detonation is based on the assumption that a spalling rock with a certain impact velocity can cause the explosives or cased charges to initiate a detonation. From all tests and analyses, the impact velocity required to initiate a detonation is generally quite high even for sensitive explosives (Table 13.1). In a series of laboratory tests reported by James and Haskins [8], the minimum impact velocity for a projectile impacting on bare TNT is 500 m/s in order to cause a detonation. For cased charges, this minimum impact velocity can go up to 1000 m/s.

Joachim and Smith used 120 m/s as a typical spall velocity required to initiate many explosives in their analysis of chamber separation to prevent propagation [10]. This is similar to a study of high-performance magazines [11], which quoted threshold impact velocities of 350 ft/s (106 m/s) for sympathetic detonation by concrete and gravel debris for sensitive items. These threshold velocities were based on tests conducted by URS Systems Corporation [12] and analyses of field certification tests of high-performance magazines.

TABLE 13.1 Summary of Threshold Impact Velocities to Initiate a Detonation [9]

Reference	Impact Velocity (m/s)	Remark
Joachim and Smith [10]	120	Quoted as threshold impact velocity for sensitive items
Malvar and Tencreto [11]	110	Threshold impact velocities for concrete and gravel debris recommended for nonpropagation walls
James and Haskins [8]	500–1000	Based on series of tests on TNT using various projectiles

Based on the aforementioned analysis, it would appear that a threshold impact velocity of 100 m/s can be adopted as the criterion for prevention of propagation by impact of a spall rock [9].

In order to initiate a detonation by impact, the explosion loading in an adjacent chamber must produce a fly rock with sufficient mass and fly velocity. A fly rock can be created by either spalling or a catastrophic failure of the rock column or roof.

For spalling, the incipient peak particle velocity (PPV) required to create a fly rock traveling at 100 m/s would be roughly 60 m/s. Such a PPV would not be realistic, as the associated stress at 60 m/s would far exceed the strength of the rock, causing a catastrophic failure. In such a case, the velocity of the fly rock would be determined by the expanding gas pressure. However, if the charge is not fully coupled and has a relatively low loading density, as is the case in ammunition storage, the gas pressure is also unlikely to cause the fly rock to travel at a very high velocity. In the underground explosive test (UET) discussed previously, the highest observed velocity of fly rocks for Zone 1 (full tunnel closure) is only about 30 m/s [13]. These tests had been done with loading densities between 248 and 1220 kg/m^3 [14].

Thus, it has been concluded by Zhou and Jenssen that at the range of loading densities typical of underground explosives storage, it is unlikely that a fly rock, either from spalling or from catastrophic failure, can cause a sympathetic detonation by impact, even at near ranges [9].

13.8 CASE HISTORY

Underground Ammunition Storage Facility, Singapore

Singapore's skyrocketing property prices are a good reflection of the land scarcity there, emphasizing the need to use land resources efficiently. This was a top priority when an underground ammunition facility (UAF) was planned in Singapore.

The facility requires only 10% of the land that a conventional aboveground depot would need. The strength of the granite formations at the quarry, complemented by safety features such as blast doors, debris traps, and expansion chambers, can contain the risks of ammunition storage without the need for a large buffer zone. The insulated environment also enables energy savings of 50%.

The facility needs 20% less manpower to operate than a conventional one because of automation and technology and will need less energy for cooling due to the natural insulation provided by granite caverns (see Figs. 13.8 and 13.9).

The Defence Science and Technology Agency (DSTA) of Singapore created new safety standards with the development of Singapore's first UAF—standards that have received international endorsement from the NATO Underground Storage Working Group.

FIGURE 13.8 View of underground ammunition storage facility in Singapore. *(source: www .innovationmagazine.com/innovation/volumes/v3n2/free/features1.shtml)*

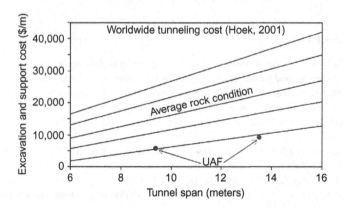

FIGURE 13.9 Tunnel excavation cost of an UAF in Singapore.

Application of DSTA criteria, coupled with innovative safety design, has resulted in a 90% reduction in the land area needing sterilization for development as safety buffers surrounding the ammunition facility—in comparison with a similar conventional aboveground storage facility. By constructing the facility underground, builders saved 300 hectares—a land area equivalent to 400 soccer fields or half the size of Singapore's Pasir Ris New Town.

Safety constitutes the most important factor in the design of the UAF. In the unlikely event of an accidental explosion, the protective design of the UAF will minimize the hazard beyond its boundaries. Many innovations in explosive storage safety, lightning protection, and fire protection enhance the protection of personnel working at the UAF. The Bukit Timah granite into which the UAF is built has six times the strength of normal concrete.

By building flexibility into the layout, the facility can be modified to meet future needs and changes. In addition, underground installations do not experience the weathering effects of rain or sun. The life-cycle costs of the installations will be lower because of reduced operating and maintenance costs, and the projected energy cost should be less than 50% of similar aboveground facilities.

The DSTA performed extensive technical studies and tests in the area of storage of explosives underground, in collaboration with overseas and local partners. In Singapore, it carried out research studies with the National University of Singapore's (NUS) Centre for Protective Technology, Nanyang Technological University's Protective Technology Research Centre, and the Institute of High Performance Computing. With NUS, research work focused mainly on the area of underground ammunition storage safety, which included numerical modeling of ground-shock propagation and response of high-rise reinforced concrete buildings.

Because of its work on the UAF, the DSTA now represents Singapore as a member of the NATO Underground Storage Working Group and the Klotz Group, entities comprising international experts responsible for undertaking technical studies and testing explosives storage safety.

REFERENCES

[1] Zhao J, Zhou YX, Hefny AM, Cai JG, Chen SG, Li HB, et al. Rock dynamics research related to cavern development for ammunition storage. Tunnel Undergr Space Technol 1999;14(4):513–26.

[2] North Atlantic Treaty Organization. Manual of NATO safety principles for the storage of military ammunition and explosives. Allied ammunition storage and transport publication, chapter 2. NATO International Staff-Defence Investment Division, AASTP-1, edition 1, Brussels; 2006. p. 431–61.

[3] Davis LK, Song S-Y. Technical managers final report. Joint U.S./ROK R&D program for new underground ammunition storage technologies TR SL-97-10 and UAST-TR-97-002 Savanna, IL: U.S. Army Technical Center for Explosive Safety; 1997.

[4] U.S. Army Corps of Engineers Drawings. Definitive drawings underground storage facility (DEF 421-80-04), Cincinnati, OH.

[5] NO(ST)(UGS/AHWP)IWP 6-98 dated 24 November 1998 one-dimensional blast wave propagation.

[6] NO(ST)(UGS/AHWP) IWP 8-98 dated 24 November 1998 model tests of accidental explosions in underground ammunition storage. II. Blast wave propagation in tunnel systems.

[7] DoD. Department of Defense ammunition and explosives safety standards. DoD 6055.9-STD, (www.ddesb.pentagon.mil); 2004. p. 264. Accessed in 2008.

[8] James HR, Haskins PJJ, Cool MD. Prompt shock initiation of cased explosives by projectile impact. Propellants Explosives Pyrotech 1996;21:251–7.

[9] Zhou Y, Jenssen A. Internal separation distances for underground explosives storage in hard rock. Tunnel Undergr Space Technol 2009;24:119–25.

[10] Joachim CE, Smith DR. WES underground magazine model tests. In: Minutes of the 23rd DoD explosives safety seminar, Department Of Defense Explosives Safetyboard Alexandria Va; 9–11 August 1988.

[11] Malvar LJ, Tancreto JE. Sympathetic detonation criteria for single concrete debris impact. Technical report TR-2043-SHR, Naval Facilities Engineering Service Center, Port Hueneme, USA; 1995.

[12] URS Systems Corporation. Investigation of explosives sensitivity to fragments and overpressure, NWC TP 4714, part 1. China Lake (CA): Naval Weapons Center; 1969.

[13] Department of the Army. Design of underground installations in rock: penetration and explosion effects. TM 5-857-4, Headquarters, Department of the Army, Washington DC; 1961.

[14] Odello RJ. Origins and implications of underground explosives storage regulations. Technical memorandum TM 51-80-14, Civil Engineering Laboratory, Naval Construction Battalion Center, Port Hueneme, USA; 1980.

[15] Hoek E. Practical rock engineering. Free download available at: www.rocscience.com/hoek/PracticalRockEngineering.asp; (cited in 2009), 2001.

Chapter 14

Underground Nuclear Waste Repositories

Intellectual growth should commence at birth and cease only at death.

Albert Einstein

14.1 INTRODUCTION

Permanent isolation or "disposal" of high-level radioactive waste is a major challenge, one that may jeopardize the future of nuclear power. It is also a challenge that must be met. We cannot simply "wish away" the volume of radioactive waste that is now in temporary storage, awaiting the development of a permanent storage facility. Political difficulties have also been added to the technical challenges. To the public, words such as "radioactive" and "nuclear" are now synonymous with "cancer," "weapons of mass destruction," and "Armageddon" [1]. Canada, Australia, and Kazakhstan are the three top producers of uranium.

Although everyone is exposed daily to radioactivity that has always been present naturally in some materials of the earth and to radioactivity produced by the sun's rays (thus combining to produce what is referred to as "background radiation"), the concentrated intense level of nuclear waste is considered to be especially evil. It is worth noting that background radiation levels vary considerably with location, depending on local factors such as ground elevation, dominant rock types, and origins of drinking water.

This fear of nuclear waste is not without some justification. Although the level of radioactivity of this waste decreases exponentially with time, some components decay very slowly, remaining at dangerous levels and concentrations for thousands—even hundreds of thousands—of years. Devising ways to ensure that the wastes are isolated for such long periods pose unusual constraints on the possible engineering designs. A number of creative solutions have been proposed in the past—among them, rocketing the waste into a solar orbit or placing it, via deep boreholes, into subduction zones at tectonic plate boundaries so that the waste is carried into the interior of the earth's crust [1].

Other options have been less extreme. The "subsea bed" option, for example, proposes to drop "torpedo-shaped" canisters of waste to penetrate deeply and be sealed into unconsolidated thick sediments on the sea floor. Should some radioactive atoms or "radionuclides" eventually escape from a canister, their rate of movement toward the sea bed would be retarded considerably by chemical adsorption and other reactions in the interstices of the sediments. Any radionuclides that reach the sea bed would be diluted to innocuous concentrations in the large volumes of sea water. Thus a repository on an island would be an excellent solution.

Today, after considerable debate over the varied scientific, technical, and sociopolitical issues surrounding each option, international consensus has developed in favor of on-land geological disposal of the waste in underground "repositories" some 500–1000 m or so below the surface. The design and location of the repository must be such that the radioactive waste will be safely isolated from the biosphere for the long times necessary to ensure that eventual releases will be extremely small both in total quantity and in level of radioactivity. Different countries have defined allowable levels in different ways, but all correspond essentially to a small fraction of the daily background radiation, much lower than the natural variations found around the globe [1].

It is very unlikely that all countries will possess the political, economical, and geological factors necessary to create permanent geological disposal programs for their materials soon or ever. For these nations, regional or international spent-fuel management solutions must remain as a viable option for both environmental and security reasons. Moreover, an international repository can be optimized for safeguarding properly the fissile materials from all users of the facility [2]. In the near future, scientists should be able to succeed in research on repositories, which are the most reliable and most durable and are extremely safe to people on the ground for thousands of years. *Nature may have created life forms whose food is radioactive waste. Further, it should be possible to change the atomic structure of radioactive waste in the future to convert them to safe materials.*

The principal route by which radionuclides could move from an underground repository to the biosphere would be by transport within groundwater flowing through the repository toward the ground surface. A low potential for such groundwater flow is an important characteristic of a potential site. Intrinsically, low hydraulic conductivity, or low "permeability," of the rock mass, preferably together with a low hydraulic driving force or hydraulic "gradient" across the site, is desirable. Thus the underground site must be dry with no groundwater table, far away from fault zones, and lie in low seismic zones. The underground opening must be in a nonsqueezing ground condition or competent rock obviously [overburden $H \ll 350Q^{1/3}$ m; where Q is rock mass quality]. The rock burst condition may be met below 900-m depths in very brittle rocks such as granite and granodiorite (Fig. 4.3). Because a tunnel boring machine (TBM) will be struck in rock burst conditions, in situ high-stress zones, as well as high thermic zones, should be avoided. Temperature increases at a rate of about 30°C per kilometer depth in addition to average ground level temperature. A small tunnel in massive hard rocks ($B<<2Q^{0.4}$ m; where B is tunnel width or diameter) may have a life span of 100,000 years, like natural caves. Moreover, a safe depth of overburden (H) should also be determined to ensure no rock burst or squeezing (so $H<<1000/B^{0.1}$ but $J_r/J_a>>1/2$). Nuclear reactors may be planned on top of the repository.

When geological isolation was first proposed in the early 1950s, location of the repository in thick salt formations was recommended. The existence of these formations, which are very soluble in water, suggested a probable lack of flowing groundwater in contact with the salt. Also, because salt has the desirable feature of flowing or "creeping" over time, excavations made to place the waste in the salt would eventually close and seal these radionuclides.

14.2 TYPES OF RADIOACTIVE NUCLEAR WASTE

Natural radioactivity is found everywhere, but "radioactive wastes" have higher radio activity levels than normal.

These wastes include spoil produced from mines; materials contaminated by the use of radionuclides in medicine, industry, and research; and contaminated and activated materials from the operation and decommissioning of nuclear power stations and from the production or dismantling of nuclear weapons (Fig. 14.1). Symbols in Figure 14.1 are explained here.

Even though the quantities are rather small, concern is usually focused on the most toxic and longest lived "spent fuel/high-level waste (SF/HLW)," which is

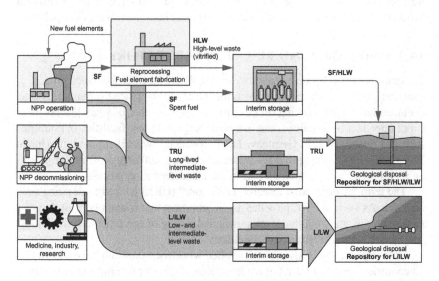

FIGURE 14.1 Different types of radioactive waste and their disposal *(source: www.grimsel.com).*

so radioactive over the first few hundred years that the resultant heat production has to be taken into account.

SF—spent fuel rods from nuclear reactors conditioned for direct disposal.
HLW—high-level waste resulting from the reprocessing of spent fuel rods (vitrified waste).
TRU or ILW—this waste is classified as intermediate-level waste (low heat production) but contains significant concentrations of long-lived radionuclides; may include reactor internals and reprocessing waste.
L/ILW—low/intermediate-level waste; very wide diversity of radioactive materials from the nuclear power industry and users of radionuclides in medicine, industry, and research. Conditioned in a solid form (usually cement) for disposal.

The used or burnt bundles of nuclear pallets and rods are generally known as "spent nuclear fuel." This spent nuclear fuel is full of plutonium and uranium, which is reprocessed to extract usable uranium and plutonium in reprocessing plants. During reprocessing, the spent nuclear fuel is dissolved in acid and undergoes different stages of processing. After the completion of reprocessing and extracting the uranium and plutonium, the liquid is left. The extracted uranium and plutonium is used again as fuel in bundles after forming pallets. The leftover liquid, however, contains long-lived radioactive waste known as "high-level radioactive waste." It includes plutonium, strontium, cesium, uranium, etc. Out of all these, strontium and cesium are highly active and main sources of heat. Efforts are being made to increase the density of this radioactive waste by extracting water to reduce the volume of nuclear waste. Liquefied HLW waste is difficult to handle. Therefore, it is solidified by mixing glass matrix or ceramics or polymer.

14.3 UNDERGROUND RESEARCH LABORATORY

To decide where to locate a repository for HLW, a lengthy and detailed process of characterizing the rock mass in which the waste will be placed is required. Some countries have been working on this process for a number of years [3]. As more experience has been gained, the use of underground test facilities in countries such as Canada, France, Germany, Japan, Sweden, Switzerland, and the United States has been found to provide a means of carrying out large-scale investigations on rock masses as an important part of the characterization process.

The term "underground research laboratory" (URL) is used to describe such facilities. As such, a URL provides an environment that is analogous to that of a repository site, allowing the development and testing of equipment, methodology, and models under fully realistic conditions.

Witherspoon and Bodvarsson compiled worldwide data and have shown that 13 countries are using URLs or are in various stages of planning and developing such facilities (Table 14.1) [3]. Two countries, Japan and Switzerland, have each developed two URLs, which enable them to carry out research investigations in

TABLE 14.1 Developments in Radioactive Waste Isolation from 2001—Third Worldwide Review [3]

Country	Lead Organization	Potential Site	Prospective Rock Type	Status of Site Characterization	Prospective Design/ Engineered Barrier	Near-Term Plans
Argentina	Commission National de Energia Atomica (CNEA)	TBD	Clay, evaporates, volcanoclastic, granite	Seven provinces chosen for site-selection purposes	To be determined (TBD)	Data being organized in GIS. Fracture mapping of satellite images is under way
Armenia	Institute of Geological Sciences of the National Academy of Sciences	Somkheto-Karabakh, Central folded zones	Volcanic, gypsum, salt and clay, granite	Geological and geophysical investigations	TBD	Complete geological, geophysical, volcanological, and seismotectonic investigations
Belarus	Institute of Geological Sciences	Polessie area, Gomel region	Granite, salt, clays	Geological and geophysical investigations of region. Numerical models of nuclide migration developed	Originally near-surface repositories; trench-type burial in future	Choose burial grounds with natural barriers, install barriers where needed, provide monitoring and testing equipment
Belgium	SCK-CEN	TBD Investigating Boom and Ieper Clay Formations	Clay	URL used 25 years for Boom Clay research, now studying effects of waste heat on safety and feasibility or clay	HLW in stainless steel overpack with bentonite backfill and concrete liner	EURIDICE Consortium managing URL research including PRACLAY project on effects of waste heat on clay

(Continued)

TABLE 14.1 Developments in Radioactive Waste Isolation from 2001—Third Worldwide Review [3]—cont'd

Country	Lead Organization	Potential Site	Prospective Rock Type	Status of Site Characterization	Prospective Design/ Engineered Barrier	Near-Term Plans
Bulgaria	Geological Institute of Bulgarian Academy of Sciences	TBD Loess near Kozludu NPP (LILW), marl at Sumer, and granite in Sakar pluton (HLW)	Loess for LILW, clayey marls and granite for HLW	Site-selection methodology developed, funding needed for specific investigations	LILW prestressed concrete cubes at NPP; HLW in deep repository	Need waste-management regulatory body and funding for geological surveys and specific investigations
Canada	Ontario Power Generation (OPG)		Granite	Since 1978, URL and other facilities used in site characterization studies and technology development	Two designs: copper containers with waste in boreholes or drifts. Buffers and backfill seal off containers and rooms	Develop waste-management organization to work with all stakeholders to develop approach that is socially, environmentally, and financially acceptable
China	Beijing Research Institute of Uranium Geology (BRIUG), China National Nuclear Corporation (CNNC)	Jiujing Block in Beishan region of Gansu Province, NW China	Monzonitic granite	Beishan region selected from geological and geophysical investigations; two boreholes drilled and cored	TBD	Comprehensive laboratory studies on core samples, investigate two other blocks in the next 5 years and plan to start URL construction by 2015

Croatia	Hazardous Waste Management Agency (APO)	Trgovska gora site in Banovina region	Clayey schist, clay, sandstone	Analysis based on existing data and reports. Mines from war prevent field investigations	Steel tube waste containers inside near-surface concrete vaults	Detailed geological mapping, geophysical surveys, and borehole drilling to provide data for site characterization.
Czech Republic	Radioactive Waste Repository Authority (RAWRA)	Dukovany for ILW and LLW. TBD for HLW	Granite for HLW	Five areas selected for site characterization. Methods and techniques to be developed at Melechor Massiv	Steel container, bentonite buffer and backfill	Two candidate sites selected by 2015 and final site by 2024. URL starts at site in 2025 and in operation by 2030
Finland	Posiva Oy	Okiluoto near NPP at Eurajoki, with existing repository for LILW and new repository site for HLW	Granite	Olkiluoto chosen from detailed site characterization, favorable EIA, and small social impact. Eurajoki OK'd site and parliament gave final approval 5/18/2001	Encapsulation in bedrock several hundred meters deep in metal canisters with buffer and backfill	Verify site suitability, define adequate repository space, characterize host rock for repository, safety assessment, and plan construction. Submit construction license application
France	Agence Nationale pourla Gestion des Dechet Radioactifs (ANDRA)	Meuse/Haute-Marne and 15 potential sites in granite areas	Argillite at Meuse site and granitic batholiths	Sinking shaft at Meuse site for URL in clay, and designing series of key experiments. Participating in five granitic experiments in foreign URLs	Investigating preliminary concepts for designs of disposal cells in clays using tunnels and caverns	Conduct research in URL at Meuse site to provide reliable data to answer key questions on performance assessment of repository in clay. Prepared feasibility report on granite site

(Continued)

TABLE 14.1 Developments in Radioactive Waste Isolation from 2001—Third Worldwide Review [3]—cont'd

Country	Lead Organization	Potential Site	Prospective Rock Type	Status of Site Characterization	Prospective Design/ Engineered Barrier	Near-Term Plans
Germany	Federal Office for Radiation Protection (BIS)/ Federal Institute for Geosciences and Natural Resources (BGR)	Gorleben salt dome, konrad iron ore mine	Rock salt, hard rock sediment within clay barrier	Licensing procedure for Konrad finished in 1992 but not yet approved. Exploration at Gorleben stopped. New procedure for site selection expected in 2002	Steel container, backfill material depending on waste type and host rock	Pursue concept of constructing one simple repository for all types of radioactive wastes. Participating in R&D programs in Belgium, France, Spain, Sweden, Switzerland, and United States
Hungary	Public Agency for Radioactive Waste Management (PURAM)	Uveghuta site for LILW, Boda site for HLW	Granite at Uveghuta site, Boda Claystone at Boda site	At Uveghuta, site characterization and social impact programs peer reviewed and approved by IAEA. HLW to be stored at NPP for up to 50 years pending policy decisions	At Uveghuta site, waste drums and disposal containers to be emplaced in tunnels with clay in backfill material	Site characterization and repository design at Uveghuta to continue, public outreach program to continue to establish long-term relationship between local communities and project management
India	Bhabha Atomic Research Centre	Sankara pluton in north-west Rajasthan	Granite	Geophysical surveys, geological and geochemical studies, hydrogeological/rock-mechanical testing carried out	TBD	Favorable results over an area of a few thousand km² indicate need for additional, more detailed investigations

Italy	National Agency for New Technology, Energy, and Environment (formerly ENEA)	Near-surface repository for LLOW. Long-term storage for HLW at the same site	TBD	Nuclear energy phased out after 1987. General site-selection process using GIS methodology, in three steps, is ongoing over entire country	Repository with modules of reinforced concrete containers and steel boxes for LLW. Long-term storage of HLW in castor-type casks	Third step in GIS methodology currently ongoing for 300 suitable areas identified in second step. Repository scheduled to begin operation soon
Japan	For implementation, the Nuclear Waste Management Organisation of Japan (NUMO); for R&D, Japan Nuclear Cycle Development Institute (JNC)	TBD	Crystalline or sedimentary rocks, URLs at Mizunami and Horonobe	Second progress report (H12) completed to demonstrate feasibility, safety, and reliability of disposal concept and to provide input for future siting and regulatory processes. Needed methodology being developed at Mizunami and Horonobe	Vitrified waste in steel overpack embedded in bentonite and emplaced either in tunnels or in vertical holes drilled from bottom of tunnels	Keep stakeholders informed on all developments via database on JNC website, provide public with virtual repository to visualize underground system. NUMO will keep public advised of potential sites, details of planned repository, and basis for final site selection. Work on new methodology continues at both URLs

(Continued)

TABLE 14.1 Developments in Radioactive Waste Isolation from 2001—Third Worldwide Review [3]—cont'd

Country	Lead Organization	Potential Site	Prospective Rock Type	Status of Site Characterization	Prospective Design/Engineered Barrier	Near-Term Plans
Korea	Korea Hydro & Nuclear Power Co. (KHNP) for LILW, Korea Atomic Energy Research Institute (KAERI) for HLW	TBD	Andesite for LILW. Mesozoic plutonic rocks for HLW	Preliminary conceptual design and safety assessment for LILW repository in rock cavern and vault options completed. Plutonic rocks screened as primary host rock for HLW repository. Conducted radionuclide migration studies and developed performance assessment code	Rock storage cavern and cement-grouted vault for LILW. Encapsulate HLW in corrosion-resistant containers in boreholes, with bentonite buffer, drilled in tunnels at depth of 500 m	Preliminary assessments of conceptual facilities for LILW provide firm foundation for site-specific assessment activities. Site-specific data required for next stage of LILW project. Next step for HLW repository is further development of repository concept using field data from in situ investigations at specific site(s)
Lithuania	Lithuanian Energy Institute	TBD for SNF and LILW. Existing solid LILW storage facilities at Ignalina NPP cannot be converted to repositories	For SNF, clay, anhydrite, salt, and crystalline bedrock	Analysis is going on, based on existing data and reports	TBD for SNF. Reference design for near-surface LILW repository (concrete vaults)	For SNF, research to develop competence in performance assessment and to select disposal concept that can be adapted to different sites; development of site-selection methodology. For LILW, siting for near-surface repository

Country	Organization	Site	Rock type	Status	HLW container	Recommendation
Netherlands	Ministry of environment Steering Commission for Research (CORA)	Retrievable disposal site for HLW at Borsele NPP	Rock salt, clay	Analysis of long-term retrievable storage at surface and underground, in either salt or clay, for up to 300 years appears technically feasible	HLW container in individual cells in tunnel wall. Crushed salt used for buffer in salt; clay–bentonite in clay	CORA recommends continuation of research program to further improve technical solutions and involve stakeholders in ethical and social aspects
Poland	National Atomic Energy Agency of Poland	Shale at Jarocin and salt domes at Damaslawek, Klodawa, and Lanieta	Triassic shale in SW Poland, salt domes in the Polish lowland	First geological review led to selection of 44 sites, of which 4 were chosen as promising: 1 in shale and 3 in salt domes	TBD	Continue more detailed site-selection process and construct URL in one of the salt domes to investigate argillaceous-salt conditions in situ
Romania	Institute of Nuclear Research	TBD	Salt	Long-term safety assessment for repository in hypothetical salt formation has been carried out	TBD	As site characterization work is carried out and in situ data become available, more realistic safety assessment investigations will be made
Russia	Minatom; All-Russia Designing and Research Institute of Production Engineering	Tomsk-7, Krasnoyarsk-26, Dimitrovgard, Mayak, Novaya Zemlya Archipelago	Sand and sandstone for liquid wastes. Hard rock for solid wastes	Liquid waste isolation has worked satisfactorily since 1963. Isolation of solidified waste in hard rock massifs, mined out areas, and permafrost now being investigated	TBD	Disposal of liquid radioactive waste to be completed by 2015 and projects will be shut down. Solidified waste will be stored at surface while geological repositories are being researched and constructed, with operation after 2025

(Continued)

TABLE 14.1 Developments in Radioactive Waste Isolation from 2001—Third Worldwide Review [3]—cont'd

Country	Lead Organization	Potential Site	Prospective Rock Type	Status of Site Characterization	Prospective Design/ Engineered Barrier	Near-Term Plans
Slovak Republic	Decom Slovakia	Tribec, Ziar, Veporske vrchy, Stolicke vrchy in granitic rocks. Cerova Vrchovina, Rimavska kotlina in sedimentary rocks	Granite, siltstones, clays	A revised program in development activities has been used to select four prospective granitic sites and two argillaceous sites. Selection of a host rock will not be made before 2005. Selection of candidate sites is expected	Proposed disposal container with seven WWER-440 SF assemblies to have outer wall of carbon-steel coated with nickel and inner wall of stainless steel. The inner cask would be an aluminum alloy	Activities should lead to proposal for a first reference disposal concept, a public involvement program, information database, investigation of prospective sites, revision of siting criteria, performance assessment based on available data, and selection of materials for engineered barrier
Slovenia	Agency for Radioactive Waste (ARAO)	TBD	Unconsolidated sediments, hard clay, granite	Areas suitable for LILW repository selected, preliminary geological assessment done	Geological conditions suitable for disposal in surface and underground	Site suitability investigations to be carried out, subject to public response. Plan to select site
South Africa	South Africa Nuclear Energy Corporation (NECSA)	Vaalputs National Radioactive Waste Disposal Facility	Clay, granite	Drilling in Vaalputs area (1996) found excellent granitic rock, but work was stopped—new national policy now being drafted	TBD	If geological disposal is to be part of national policy, all stakeholders are to be involved. International cooperation is essential, various options (including regional repository) to be included

Spain	Empresa Nacional de Residuos Radioactivos (ENRESA)	TBD	Clay, granite	Detailed analyses of several potential repository sites have been made. Large database from this work now being managed on EIS, which will be updated	Carbon-steel canisters embedded horizontally in bentonite buffer spaced 2 m apart in drifts	Develop geological disposal R&D program with models for site characterization, flow and transport scenarios, and performance assessment. Develop generic design for repository in clay or granite; study natural analogs; establish safety criteria
Sweden	Swedish Nuclear Fuel and Waste Management Co. (SKB)	Oskarshamm in SE Sweden, Tierp and Osthammar in northern Uppland	Granite URL in granite at Aspo	Feasibility studies led to three potentially suitable sites for deep repository, approval for site investigations needed from local municipalities. URL at Aspo conducting R&D on methodology needed for deep repository	Waste in copper canister with cast inserts embedded in bentonite in vertical holes in tunnel filled with bentonite and crushed rocks at depth of 500 m	With local approval, site investigations started. Much work on canister fabrication under way at Canister Laboratory in Oskarshamm. Research on repository technology now concerned with a retrieval test, backfill and plug test, and prototype repository
Switzerland	National Cooperative for the Disposal of Radioactive Waste (NAGRA); Genossenschaft fur nukleare Entsorgung Wellenberg (GNW)	Wellenberg for LILW, HLW/SF/TRU siting studies in northern Switzerland	Marl at Wellenberg; clay, granitic basement in northern Switzerland; URLs at Mt. Terri in Opalinus Clay and at Grimsel in granite/granodiorite	Wellenberg LILW site accepted at local level but blocked at cantonal level by narrow margin. Opalinus Clay and granitic basement in northern Switzerland investigated extensively, including three-dimensional seismics and deep boreholes	SF/HLW in steel canisters embedded horizontally in tunnels with bentonite backfill. LILW/TRU in concrete emplacement containers in caverns backfilled with cementitious grout	Second referendum at Wellenberg in 2002 permitted exploration tunnel to gather data to support application for construction license. Siting feasibility project for HLW/SF/TRU focused on the Opalinus Clay of the Zuercher Weinland was produced in 2002

(Continued)

TABLE 14.1 Developments in Radioactive Waste Isolation from 2001—Third Worldwide Review [3]—cont'd

Country	Lead Organization	Potential Site	Prospective Rock Type	Status of Site Characterization	Prospective Design/Engineered Barrier	Near-Term Plans
Taiwan	Fuel Cycle and Materials Administration (AEC)	Little Chiu Yu at Wu-Chiu Hsiang for LLRW. SF currently stored in onsite pools at NPPs.	Granite, shale, mudstone	EIS for Little Cdhiu Yu under review with Taiwan EPA. Approval of feasibility and safety analysis reports plus EIS needed for final approval of site	TBD	SF disposal under study in project spanning 40 years (1991–2031). Expect SF disposal site identified by 2016 and repository commissioned by 2032. On-site dry storage to supplement on-site pool storage.
Ukraine	Institute of Geological Sciences	Korosten pluton and Malakhov block in Ukrainian shield and salt domes in Dnieper-Donets depression	Granite, salt domes	Site selection and characterization methodologies defined, funding problems with economic restrictions	TBD	Complete R&D on site characterization (1999–2005), characterize selected site, develop URL, demonstrate site safety, obtain license and decision on construction (2005–2020)

United Kingdom	United Kingdom Nirex Ltd.	TBD	TBD	Request for permission to build a URL near Sellafield rejected by local council and decision supported by secretary of state for the environment. Work at Sellafield terminated	TBD	Parliamentary review (1999) points to need for public acceptance of policy on waste management before problem can be settled. Citizen's panel issues number of suggestions. Government issues proposal to develop, and implement, a waste-management program that inspires public support and confidence
United States	U.S. Department of Energy	Yucca Mountain, Nevada	Volcanic tuff	Site selection and site characterization methodologies have been developed and applied evaluating Yucca Mountain	Waste within two concentric cylinders (stainless steel inside corrosion-resistant nickel alloy) covered with drip shield and placed horizontally in drifts	Quantitative assessments of long-term performance of repository for various features, events, and processes are going on. Performance-conformation program established to monitor and confirm if repository is behaving as expected. These preclosure-period activities may last up to 300 years

Note: TBD - To Be Determined

two different rock systems at the same time. This increasing use of URLs has led to another important development in which two or more countries can contribute to joint projects in underground research on problems of mutual interest. At present, argillites and granites are the dominant rock types being investigated in the URLs of Europe, and a number of joint projects have been set up to investigate the characteristics of these two different rock types.

For example, in Switzerland, Nagra (National Cooperative for the Disposal of Radioactive Waste) and 17 other organizations from nine countries have been involved in research on granite of the Swiss Alps at 500 m depth below ground in the underground research laboratory at the Grimsel test site (GTS). In effect, over a period of more than 20 years the GTS has become an international center for in situ research on radioactive waste isolation in granite.

14.3.1 Stripa Underground Research Facility

In 1977, the U.S. Department of Energy and SKB, the Swedish Agency for the management of nuclear waste, entered into a cooperative arrangement to pursue investigations at the Stripa mine in granites in Sweden. A Swedish–American cooperative (SAC) project was carried out between 1977 and 1980 at the Stripa mine. The SAC project consisted mainly of the three following elements [4]:

1. Heater experiments
2. Assessment of fracture hydrology
3. Geophysical measurement

Extensive information was obtained on the mechanical response of the granular rock to heat load and on the groundwater flow characteristics in fractured crystalline rock.

The research is carried out as an autonomous project under the sponsorship of the Organisation for Economic Co-operation and Development/Nuclear Energy Agency and is managed by SKB. Participating countries are Canada, Finland, France, Japan, Spain, Sweden, Switzerland, the United Kingdom, and the United States.

The Stripa project includes a number of subprojects with different objectives, budgets, and time schedules. Research is divided into the following areas:

- Detection and mapping of fracture zones
- Groundwater characteristics and nuclide migration
- Bentonite clay for back-filling and sealing (engineered barriers)

Crystalline rock formations, such as granite and basalt, provide more than adequate shielding against radiation and will also disperse the heat produced by radioactive waste. Deep underground disposal of radioactive wastes in rock formations with no known valuable minerals also makes future intrusion by humans less likely. *Geological disposal of radioactive waste is an entirely passive system that does not depend on human involvement to ensure safety over very long time periods.*

No hazard to humans or the environment will occur while the waste remains contained in an appropriately sited repository.

Because of the long time periods during which radioactive wastes remain radioactive, the possibility of release of radionuclides from the repository and their transport to the biosphere must be assessed. The only possible release mechanisms are dissolution and transport in a flowing groundwater. Therefore, extensive studies of groundwater–rock-engineered barrier interactions have to be undertaken. We know that water seeps mostly through fractures depending on stress conditions and temperature. *We need to study the effect of earthquakes on permeability.*

Development of methods and techniques for such studies and verification of previously obtained laboratory results by in situ experiments are the general objectives of the Stripa project. In addition to research at the Stripa mine, Sweden has a very well-developed program for the disposal of different radioactive wastes [4].

14.4 CONCEPT OF BARRIERS

The basic criterion accepted worldwide in the design of a repository for high-level waste is the "multibarrier principle," which holds that the repository system should be composed of different redundant natural (geological) and engineered (multiple defense system) barriers. Performance of the repository is dependent on the total system rather than the performance of a single barrier. Therefore, in a comparison of different repository systems, the evaluation of the long-term performance and safety of each individual barrier is combined with an evaluation of the degree of dependence that the overall safety has on the performance of that particular barrier.

A repository system may include a barrier with a very high level of safety; however, if overall safety becomes low if that barrier fails, the system is not as robust as intended. The barriers, in general, are natural and engineered as follows [5]:

Engineered barrier system (EBS)
- Spent fuel itself, which has a low solubility in a reducing environment
- Canister containing the fuel
- Bentonite clay buffer surrounding the canister

Natural (geological) barrier
- Bedrock, divided into near field and far field

Studies are focused on the natural system attributes that optimize EBS performance, including relative tectonic stability, low groundwater flux, favorable geochemistry, and a low risk of disruptive events.

A safety analysis [6] concludes that the low solubility of spent fuel, the canister (copper canister filled with cast lead), and the bentonite buffer efficiently isolate the radioactive inventory. Radioactive fission products and all actinides with high initial activity remain in the engineered barriers. Cesium (^{137}Cs) and strontium (^{90}Sr) decay to low levels before water may come in contact with the fuel in a canister with a hole in the copper shell. Solubility limits and sorption in the bentonite prevent other specimens with high initial activity, such as the actinides plutonium (Pu), neptunium (Np), and americium (Am) and the long-lived fission products zirconium (^{93}Zr), paladium (^{107}Pa), and tin (^{126}Sn), from reaching the near field rock. In principle, only the highly soluble carbon (^{14}C), iodine (^{129}I), and cesium (^{135}Cs) and the long-lived uranium daughters radium (^{226}Ra) and potactinium (^{231}Pa) may migrate out into the rock mass [5].

Bentonite is an excellent sealant material for containment of radioactive waste due to its swelling and self-healing characteristics, low permeability (7×10^{-8} cm/s), sorption qualities, and longevity nature. Temperature has no significant influence on the sealing properties of bentonite/crushed tuff plugs up to 60°C. There is no possibility of piping due to flow of water radially [7], as swelling pressure will seal drill holes when seepage occurs. The in situ permeability of in situ bentonite plugs is slightly higher. The properties of commercial bentonites are as follows [8]:

Cohesion	93 kPa
Angle of internal friction	24°
Residual cohesion	30 kPa
Residual angle of internal friction	10°
Shrinkage limit	16–38
Plastic limit	33–54
Liquid limit	340–693
Specific gravity	2.64–2.82
Proctor's density at optimum moisture content	1142–1525 kg/m^3
Laboratory permeability	7×10^{-8} cm/s [7]
Swelling pressure	$\cong$10 MPa

A careful and well-planned site selection process for a repository should provide a robust and safe site, satisfying the following requirements [9]:

- Specifying the existence of geological environments that are stable over appropriate timescales and provide favorable conditions for EBS performance and for the radionuclide retardation function of the geosphere
- Illustrating an appropriate design for containment and retardation of radionuclides in the EBS for a wide range of geological environments

14.5 DESIGN ASPECTS OF UNDERGROUND REPOSITORY

14.5.1 Vertical Emplacement in a Pit

Layout

The design of a repository layout with a vertical disposal system is shown in Figure 14.2. After about 40 years of interim storage, spent fuel assemblies are conditioned in copper canisters, two designs of which are studied:

- Steel canister with a 50-ram-thick copper shell
- Copper canister with an interior completely filled with cast lead

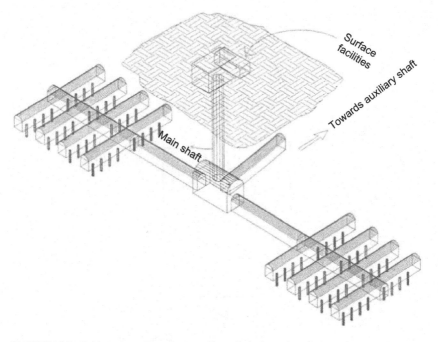

FIGURE 14.2   Design of repository layout with vertical pit emplacement system.

The canisters are emplaced in vertical positions in holes/pits bored in the floor of a gallery of drifts. One canister is placed in each hole/pit. The canister is totally surrounded by a highly compacted bentonite buffer with low hydraulic conductivity and good sorption properties.

After emplacement of the canisters, the disposal drift is backfilled with a mixture of bentonite and quartz sand (average 15% bentonite). Thereafter, each drift is plugged in such a way that axial groundwater transportation paths are cut off. Access ramps and shafts eventually are plugged and backfilled as well.

The more spent fuel in the canister, the larger the size of the canister and therefore the more heat output. Typical dimensions of the copper/steel canister (reference design) and the disposal hole being studied in Sweden are shown in Figure 14.3.

A key parameter guiding the exact design of the repository is the maximum temperature to which the bentonite can be exposed during the initial thermal pulse without altering the properties of long-term performance negatively. In laboratory experiments, it has been shown that minimum alterations take place at temperatures below 130°C. In order to leave some margin, the design temperature is set at 100°C. Maximum ambient rock temperature at a 500-m depth in Sweden is 18°C [5].

Given these temperature data, canister dimensions according to Figure 14.3, and experienced thermal conductivity of dry bentonite in deposition holes,

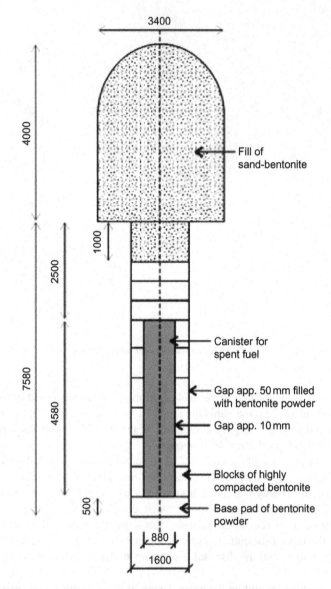

FIGURE 14.3　Disposal hole with copper/steel canister and buffer material [5].

canisters need to be spaced at 6-m intervals and disposal drifts at 25-m intervals. The temperature in the rock walls of the disposal holes as reported by Pusch and Svemar [5] reaches a maximum of about 75°C after 50 years. The maximum temperature in the bentonite/canister interface is attained about 15 years after disposal.

Disturbed Zone

The rationale for choosing to place canisters in holes in the floor was, first, to locate the waste in as intact rock as possible in order to facilitate the description of the hydraulic regime around the canister in the long run. In the 1970s, it was understood that the zone closest to a blasted drift was affected up to 2 m by explosives and that an increased hydraulic conductivity might be expected. However, the properties of this zone are better known today, and the effects of excavation on hydraulic conductivity can be estimated [5,10]. Blasting and stress relief cause fractures all round a tunnel in a rock mass, which act as "super conducting tubes" along which groundwater is transported rapidly throughout repositories. Pusch recommended Na-bentonite to make the rock mass nearly impermeable [10]. It is important to know that long-lived nuclear waste does not dissolve in water and that rock mass will also act as an effective filter and will dispose and dilute sediments of long-lived waste. However, short-lived nuclear waste can dissolve into ground water and reach people on the ground [11].

Water Inflow during Disposal

The disposal operation is performed in the following sequence:

1. Bentonite (in the form of highly compacted blocks) is installed
2. Canister is installed
3. Bentonite blocks are placed on top of the canister
4. Disposal drifts are backfilled

Under normal conditions, it takes less than one shift to emplace bentonite and a canister in one hole. The drift is backfilled later, after all of the holes have been filled. However, if any adverse event—a machine breakdown, for example—occurs during emplacement of the canister or the top blocks, more time is required. The determining factor most probably will be swelling of the bentonite blocks, resulting from water seepage into the hole. Bentonite surfaces exposed to dripping water start to swell immediately. The alteration, however, is very shallow and the "skin" can be swept off when the disposal operation starts up again [5].

A more severe problem can occur if too much water flows into the hole. Tolerance between the blocks and the rock wall (50 mm wide before backfilling with bentonite powder) is approximately 1000 liters. Tested methods of clay grouting [12] have the potential to decrease the water inflow to 1–2 liter/hr into the hole. This would mean a filling time of 3 to 6 weeks. Bhasin and Olsson have analyzed underground nuclear waste repositories using UDEC-BB and NAPSAC software, taking into account the interaction of joint seepage and stresses [13]. A test is also described. They recommend grouting of the rock mass around tunnels with superfine cement, water, and superplasticizer.

Water Flow after Repository Closure

The initial period after closure is characterized by transient temperature and water pressure conditions. The bentonite is saturated and homogenized. Eventually it obtains saturated pressure conditions (5–10 MPa). Water pressure is built up in the rock. The thermal pulse induces displacements of the rock and shear in weak areas. After several thousand years, the pulse has decayed and the rock has recovered ambient temperature.

One important function of the near-field rock during the course of these events and afterward is to provide the bentonite with a satisfactory environment that limits exposure of the buffer to clay-altering substances transported in water [5].

14.5.2 Alternative System Layout for Very Long Hole (VLH)

Layout

The VLH system design is shown in Figure 14.4. The canister is placed in the center of a full-face bored drift (horizontal) and surrounded by compacted bentonite. The dimensions have been established with respect to the basic criterion of not exceeding 100°C in the bentonite and the wish to have a high thermal load per canister. This has resulted in a choice of thermal load that is twice the load of the vertical emplacement size canister. The subsequent outer surface of the canister is

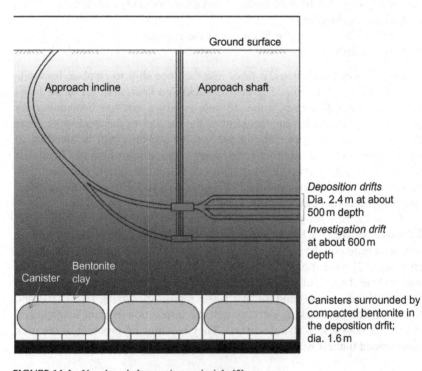

FIGURE 14.4 Very long hole repository principle [5].

proportional to the thermal load, provided that all other parameters are maintained the same. Although minor geometric differences are introduced, such as a bentonite buffer thickness of 0.4 m—instead of the 0.35 m in the vertical emplacement design—the resulting dimensions are a canister diameter of 1.6 m and a drift diameter of 2.4 m. Canisters in the base case are equipped with hemispherical ends and are 5.9 m long. They are assumed to be deposited in a row in the drift, with a center-to-center distance of 6.0 m. The length of the drift can be decided depending on the canisters to be buried and the overall thermal load accepted by the ground/rock.

With a distance of about 100 m between drifts, the neighboring drift does not affect the maximum temperature around a canister. Although the peak for this design also is reached after about 15 years in the canister/bentonite interface, the temperature drops more rapidly [14]. Concrete and shotcrete will not be used for long-term purposes obviously.

Disturbed Zone

By using the TBM technique for excavation, disturbance to rock close to the drift is minimal. Several analyses on this subject have been performed by the SKB agency [15,16].

Water Inflow during Disposal

The proposed disposal sequence is divided into three separate activities [17]:

1. The bottom bentonite bed is placed
2. The canister is placed on top of this bed
3. The remaining blocks are installed

Each sequence is separated by the exit of one vehicle that has left its cargo and the entrance of another vehicle to take care of unexpected events. The VLH system also is sensitive to water inflow during the disposal operation. An acceptable inflow of 10 l/hr along the entire canister has been estimated [14]. It is considered feasible to achieve this flow by applying the same clay-grouting method as that suggested for vertical emplacement disposal holes.

Water Flow after Repository Closure

The sequence of evolution in the near field is the same as that described for the KBS-3 system. The postulated performance of the clay barrier is also very similar. The question is whether the horizontal design provides a significantly different basis than KBS-3 for long-term isolation of the encapsulated spent fuel.

14.6 INSTRUMENTATION

Observations of the entire repository are needed for several hundred years, such as with the leaning Tower of Pisa. The following observations are as important as they are in all major important underground structures.

- Monitor change in temperature
- Monitor displacements inside rock mass
- Monitor tilt of drill holes and canister, etc.
- Monitor flow of groundwater very accurately, pore water pressure, pH values, etc.
- Monitor development of swelling pressure in bentonite with time
- Monitor radioactivity on the ground surface and outlets of underground openings especially

14.7 RETRIEVABILITY OF CANISTER

The repository shall be planned and designed in such a way that it is possible to retrieve deposited canisters. Hence, methods for retrieving the canister must also be developed and tested. The main goal for this test is to develop a method for freeing a canister from water-saturated bentonite. Such a test is being carried out and planned in some countries such as Sweden and France [18].

14.8 PUBLIC ACCEPTANCE OF RADIOACTIVE WASTE REPOSITORY

Public acceptance of the management of a radioactive waste isolation project is an important issue, and it should be addressed in the best possible manner. In approaching this problem of public acceptance or to generate public confidence in geological disposal, the Japanese have taken an unusual step to promote public understanding of the nature of an underground repository. The Japan Nuclear Cycle Development Institute (JNC) has developed a special demonstration tool named "Geofuture21" that is in operation in the JNC Tokai center. In Geofuture21, people can enter a virtual repository placed 1000 m underground by a combination of scientific simulation, three-dimensional visualization, and a motion system. In this way, the public can experience a simulated earthquake deep underground, observe the behavior of an engineered barrier system, and witness how radionuclides move through bentonite. According to [9], about 90% of over 12,000 visitors so far have responded that they could understand geological disposal quite well.

Keeping the local public informed of developments in field investigations is very important. Assuring transparency and traceability is also critical [3].

REFERENCES

[1] Fairhurst C. Geological isolation of radioactive waste: a challenge that must be met. Tunnel Undergr Space Technol 1993;8(3):313–4.
[2] Pellaud B, McCombie C. International repositories for radioactive waste and spent nuclear fuel. In: Proceedings of the INMM annual meeting. Institute of Nuclear Materials Management, New Orleans, Louisiana; 2000.

[3] Witherspoon PA, Bodvarsson GS. Introduction to geological challenges in radioactive waste isolation: third worldwide review. In: Geological challenges in radioactive waste isolation: third worldwide review (Editors: PA Witherspoon and GS Bodvarsson), chapter 1, LBNL-49767. University of California, Berkeley; 2001. p. 1–13.

[4] Morfeldt C-O. Underground construction on engineering geological terms: a fundamental necessity for the function of metropolitan environments and man's survival. Eng Geol 1991;30:13–57.

[5] Pusch R, Svemar C. Influence of rock properties on selection of design for a spent nuclear fuel repository. Tunnel Undergr Space Technol 1993;8(3):345–56.

[6] SKB. Final disposal of spent nuclear fuel, SKB 91: importance of the bedrock for safety. SKB technical report 92-20. Stockholm: SKB; 1992.

[7] Ouyang S, Daemen JJK. Sealing performance of bentonite/crushed rock borehole lugs. Dept. of Mining and Geological Engineering, University of Arizona, NUREG/CR-5685; 1992. p. 315.

[8] Swayer WD, Daemen JJK. Experimental assessment of the sealing performances of bentonite plugs. Dept. of Mining and Geological Engineering, University of Arizona, Tucson, NUREG/CR-4995; 1987. p. 291.

[9] Masuda S, Kawata T. The Japanese high-level radioactive waste disposal program. In: Witherspoon PA, Bodvarsson GS, editors. Geological challenges in radioactive waste isolation: third worldwide review, chapter 17, LBNL-49767. University of California, Berkeley; 2001. p. 167–81.

[10] Pusch R. Rock-backfill interaction in radwaste repositories. In: Hudson JA, Hoek E, editors. Comprehensive rock engineering, vol. 5. New York: Pergamon Press; 1993. p. 565–81.

[11] Bjurstrom S. Technology exists, knowledge exists, when and how will society make use of it? In: Proceedings of international conference: underground construction in modern infrastructure. Stockholm: A.A. Balkema; 1998. p. 283–8.

[12] Pusch R. Executive summary and general conclusions of the rock sealing project. Stripa project technical report. Swedish Nuclear Fuel and Waste Management Co (SKB), Stockholm; 1992.

[13] Bhasin R, Olsson R. Rock joint and rock mass characterization for nuclear waste repositories. J Rock Mech Tunnel Technol 2010;16(2):97–112.

[14] Sandstedt H, Wichmonn C, Pusch R, Borgesson L, Ionnerberg B. Storage of nuclear waste in long boreholes. SKB technical report 91-35. Stockholm: SKB; 1991.

[15] Pusch R, Hokmark H. Characterization of nearfield rock: a basis for comparison of repository concepts. SKB technical report 92-06. Stockholm: SKB; 1991.

[16] Winberg A. The role of the disturbed rock zone in radioactive waste repository safety and performance assessment: a topical discussion and international overview. SKB technical report 91-25. Stockholm: SKB; 1991.

[17] Henttonen V, Suikki M. Equipment for deployment of canisters with spent nuclear fuel and bentonite buffer in horizontal holes. SKB technical report 92-16. Stockholm: SKB; 1992.

[18] Lundqvist B. The Swedish program for spent-fuel management. In: Witherspoon PA, Bodvarsson GS, editors. Geological challenges in radioactive waste isolation: third worldwide review, chapter 28, LBNL-49767. University of California, Berkeley; 2001. p. 259–68.

Chapter 15

Contractual Risk Sharing

Engineers have to take a calculated risk, persons become wiser after an accident. If they were really wise, it was their duty to point out mistakes in the design to engineers.

Karl Terzaghi

15.1 THE RISK

Risk is a basic element of life. Calculated risk makes us bold. Life without calculated risk is inconceivable and undesirable. If all risks were eliminated, the construction industry would cease to evolve. With the hope of greater profit, the contractor develops a new method, accepting the risk that the method may not result in loss of anticipated profit. Without this element of risk he would lack the initiative and the incentive to select new techniques or execute them. The owner also assumes some risk when sponsoring a project, which may result in nullifying the projected benefits. It is well known. No risk! No gain!

While many risks in construction are inevitable, not all are. Careful, thorough, and detailed planning and engineering analyses will identify most of them, and ways may be devised for avoiding some (the known) and lessening the trauma from those that are expected but cannot be foreseen clearly enough to avoid completely (unknowns, inherent uncertainties). The greatest need for sharing of risks is for occurrences that are not expected. The execution of these plans through the construction phase must be directed toward decisive action that will meet the planned objectives, including the management of uncertainties and the risks. Contingency plans and methods for managing risks must be kept up to date and revised to meet the actual situations/hazards that arise, in consultation with all contractors.

Earlier, risks in tunneling were smaller and lesser and could be more easily classified and borne or handled by the various participants in a more equitable manner. However, today's huge complex and imbalanced risks cannot be borne solely by one of the partners. Hence, means for allocating and sharing these risks should be evolved for the common good of the project construction organizations and its beneficiaries. The actual size and probability of the risk involving cost, time, credibility, reputation, and ability to perform are unknown but real. The existence and impact of these risks should be appreciated and means of mutual benefit found. Most risks are evaluated, minimized, or eliminated, and the cost of doing this should be compared with an assessment of the original risk.

Underground Infrastructures

In too many instances the scramble to avoid risks by throwing them back and forth has made the construction scene a battle field for lawyers rather than an opportunity to accomplish useful and lasting works. The interests of the construction industry would be well served if more attention was directed to creating a construction team composed of owners, engineers, contractors, geologists, and insurers, each contributing their special expertise to solve common problems in the underground construction and each sharing risks related to their capabilities [1].

Contributions have been made by many countries toward the object of defining the sources of risk in a contract and in establishing how best, in the interest of the common good, these are shared among the parties concerned. The latter, sharing of risks, which builds upon practice in the United Kingdom and is largely accepted in Austria and other European countries, has a number of essential features, the most important of which are:

• Generally attribute acceptance of risk to the party best able to control its incidence (contractor) or, for minor risk, to make reasonable provision for its cost.
• Provide appropriate encouragement to use methods of construction that show best prospects, in the available knowledge at any time, of an economic result.
• Provide appropriate flexibility for change in construction methods to follow the range of variation in ground and other conditions foreseeable by a knowledgeable engineer.
• Simple and equitable arrangements for disposal of disputes.

The U.S. National Committee on Tunneling Technology has given recommendations on better contracting for underground construction, which include:

• Sharing of risks and their costs between the owner and the contractor. The risks are both construction risks and financial risks.
• Handling of claims is required to be expedited.
• Innovation in construction should be stimulated.
• The award of work to the qualified contractor should be assured.
• Cost savings by other means should be realized [2].

The need for better management and better contracting in underground construction in the United States has been elaborated by [3]. Contracting practices for tunneling have been discussed in detail by [4]. Muirwood and Sauer describe (i) the managerial principles for economic tunneling, resulting in a less expensive, faster, and more reliable project to the owner; (ii) greater scope for the ingenuity of the engineer; and (iii) the contractor with greater confidence for a fair return for his skill and resources [5].

Contracting practices in European countries and the United States are compared by [6]. He concluded that a successful contract for both owner and contractor is the product of a marriage between good contracting practices and good management organization.

Samelson and Borcherding [7] examined several barriers to productivity described by foremen from five different construction sites as:

- Waiting for decisions
- Waiting for materials and tools
- Rework

In Japan [8], decisions are made by a consensus approach. Although it takes time to achieve a consensus, once achieved, it assures total commitment to the successful outcome of the decision, and implementation is almost assured. It is heartening to know that some corporations act like parents to their workers and their beloved children. May God prosper their love!

The International Tunnelling Association (ITA) Working Group on Contractual Sharing of Risks, in cooperation with the International Federation of Consulting Engineers, is preparing a standard contract for tunneling work. The assessment of risk and its sharing in tunneling has been brought out by [9]. He discussed three categories of risks—functional, structural, and contractual—and how they relate specifically to the design and construction of underground openings. He stressed an urgent need for improved methods of risk assessment because the causes of functional and structural failures are complex and often interrelated. He proposed a number of recommendations concerning risk assessment.

Equitable sharing of risks means that the party bearing a greater part of the risk should be entitled to a greater share of the benefits or profits and the other parties should have no objection to it. If the sharing of risks is not equitable, then there would be an imbalance between the risks actually borne and the profits made by the parties, which may lead to disputes or litigation and consequently to delays and a higher project cost.

By including a clause covering adjustment in unit price for unknown conditions, the contractor is not tempted to escalate his item rates to cover the risk of adverse underground conditions. Full disclosure of all subsurface data available with the owner/goverment department to the tenderers/bidders may lead to a lower contract cost.

Disclaimer clauses relieving the owner of responsibility for the accuracy of underground data furnished should be deleted. If disclaimer clauses cannot be eliminated completely from a contract, at least their number should be reduced to minimize malpractice. Absence of a "changed conditions" provision in a contract will induce the contractor to put a contingency amount in his bid. As a result, incorporation of this clause is beneficial to the owner [6]. The ITA recommends that a changed condition clause be incorporated in all tunneling contracts.

Goverment departments should seek bids only from contractors having rigorous technical and financial prequalification. It has now been realized that prequalification of bidders is as much a part of construction as selecting a suitable contractor. The practice of calling prequalification tenders by prospective bidders is being adopted in new projects. Authority to settle claims, commensurate with the scope of the project, should be delegated to both the representatives of the owner and the contractors in the field. The decision of whether to use wrap-up insurance should remain with the owner.

A few subclauses under the changed conditions clause as described here are suggested for inclusion in tender/contract documents of tunneling contracts globally if they have not been considered by the owner.

- "The contractor shall promptly, and before such conditions are disturbed, notify the engineer-in-charge in writing of: (i) subsurface or latent physical conditions at the site differing materially from those indicated in the contract, or (ii) unknown physical conditions at the site, of an unusual nature, differing materially from those ordinarily encountered and generally recognized as inherent in work of the character provided for in this contract. The engineer-in-charge shall promptly investigate the conditions, and if he finds that such conditions do materially so differ and cause an increase or decrease in the contractor's cost of, or the time required for, performance of any part of the work under this contract, whether or not changed as a result of such conditions, an equitable adjustment shall be made and the contract modified in writing accordingly."

- "No claim of the contractor under this clause shall be allowed unless the contractor has given the notice required in (a) above, provided, however, the time prescribed therefore may be extended by the government or the agency executing the contract."

- "No claim by the contractor for an equitable adjustment hereunder shall be allowed if asserted after final payment under this contract."

The Norwegian practice of risk sharing in tunneling contracts has proved successful in that 80% of their proposed 2600 km of tunnels have been driven with equivalent time risk sharing built into the contracts. No disputes with relevance to changed ground conditions have been reported in the period after the risk sharing provisions were accepted in their contracts [10].

Sharing of risks in tunneling contracts and management of risks have been discussed by [11,12]. A survey of opinions of tunneling experts in Himalayan projects indicated a low priority to sharing of risks, whereas the ITA has realized its importance and brought out recommendations on sharing of risks. Crisis decision analysis is encouraged at the project site. Design of a contract document is needed for underground space technology.

Risk is defined as "the possibility of loss, injury, disadvantage, or destruction," that is, risk is an adverse chance. It is necessary to have information as to know how problems arise and with whom, what is the nature of the risks, and how to alleviate them. Risks that are either undefined or unrecognized prior to the award of a contract cause much grief later. Owners or government departments should realize that a fair contract with equitable sharing of the risks according to the ability to assess and manage them would lead to earlier completion dates at lesser costs. Current contracting practices lead many tunneling projects to wind up with tremendous increases in estimated cost, financial disasters, disputes, and litigation. The situation is aggravated by energy crises, economic uncertainty, terrorism, and shortage of materials and equipment. At the same time, if our industries are to develop their maximum technological potential, we must employ contracting practices that will encourage development.

Risks in underground construction are related to a number of factors listed here:

- Acts of God
- Accidents
- Acceleration or suspension of work
- Agencies involved
- Allocation principles of risks
- Costs
- Construction and construction failure
- Contract
- Contractor/owner inherent
- Changed conditions
- Defective design/work
- Decisions
- Delays
- Data
- Disclosures of information
- Disclaimers
- Design of supports
- Deductions
- Economic disasters
- Environmental conditions
- Evaluation
- Escalation
- Equipment
- Funding and financial failure
- Groundwater
- Individual capabilities
- Inflation
- Innovation
- Information
- Insurance
- Investigation
- Labor
- Materials
- Management

- Managerial competence
- Physical risks
- Political and social conditions
- Public disorder
- Planning and scheduling
- Pilot works
- Quantity variations
- Regulations
- Reimbursements
- Resolving problems
- Responsibilities
- Site access
- Subsurface conditions
- Subcontractor failure
- Shared risks
- Sociological problems
- Support systems
- Third-party delays
- Union strike
- Water problems

15.2 MANAGEMENT OF RISK

Timely release of funds by the governments/departments would serve as a morale booster for the contractor. A delay in running payments to the contractor affects the workmen, which certainly tell upon their efficiency. This is especially true in dishonest poor societies. A total commitment of executives increases the confidence of contractors and reduces accidents. If payment to a contractor is made quickly, tunneling will be faster naturally, due to quick reinvestment.

The subject of risk involves responsibility, liability, and accountability. The basic principle of risk relationships is that the party taking the risk (contractor) should assume the liability and either suffer the consequences or reap the benefits therefrom, depending on the outcome of the endeavor [13]. An instrumentation program to probe strata in advance of tunneling in poor rocks and to study the adequacy of supports would result in safety and economy.

Team spirit is very much lacking in poor countries in government departments because of lack of mixing of top executives among junior staff and workers and a rigid hierarchy. Very few top officers associate themselves with subordinates and their problems. *The spirit of mutual trust and benefit is very important in the risk management.*

In long tunnels, a number of contracts should be awarded for different reaches and lengths of tunnels to introduce an element of fair competition and to encourage better performance. Another factor—energy management—has nowadays become quite important in view of the monopoly of oil-rich nations. Moreover, these days, top security should be provided to all engineers and contractors against terrorists.

Figure 15.1 represents risks and risk sharing in tunneling contracts. Risks in tunneling contracts are related to the 19 factors recommended by the ITA. Risks inherent in these factors should be shared equitably among the contractor, owner, engineer, geologist, and insurer.

Engineers should be bold and try taking risks of new technologies (of internationally reputed corporations). The contract should include (i) a clause for

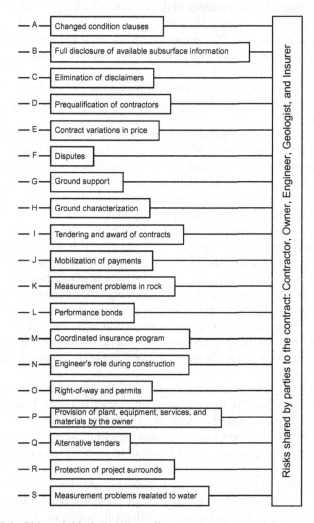

FIGURE 15.1   Risks and risk sharing in tunneling contracts.

compensation to a contractor for any unexpected geological conditions or surprises, (ii) a clause on innovations by contractors and engineers on the basis of mutual agreements, (iii) clauses for first and second contingency plans for preparedness, and (iv) a penalty for delays in construction. Obviously, a contract is not a license for injustice to any party. An injustice done should be corrected soon. The principles of crisis management are that one should not be panicky and one should not spread panic. The right persons at the right places contribute to the success of projects (according to Dr. V.M. Sharma). Team work is the key in achieving safety goals. Delay is better than error according to Thomas Jefferson.

The experience is that a number of disputes increased rapidly, as the number of clauses increased in a contract beyond a certain limiting number. According to Professor J. J. K. Daemen: "There was increase in laws after every underground disaster but it was counter-productive. A judicious liberty is essential for increasing efficiency of an organization."

Our organizing ability is increasing automatically and cyclically with time due to the law of negative bioentropy [14]. Conflicts are beneficial in increasing our inner strength.

Figure 15.2 is a conceptual model of risk sharing. Clauses or provisions in a tender/contract document will either benefit or affect adversely the interests of the persons involved in any underground construction, namely, contractor, owner, engineer, geologist, and insurer. The parties (involved) share the risks inherent in underground construction in different proportions. For equitable sharing of risks, the party taking the greater portion of the risk(s) should be entitled to a greater share of the benefits or profits due to increased costs. If profits to a party are not commensurate with the amount of risk taken by it, it will be an inequitable sharing of risks. Badrinath has developed an expert system considering the five parties mentioned earlier [15]. This software reveals whether a tunneling contract has the risks of the project shared justly and fairly among the owner, engineer, contractor, insurer, and geologist responsible for execution of the project. This expert system may be used to educate construction engineers and managers on how to improve contract documents for the mutual benefit of all concerned. Mutual benefit and trust may decrease the heavy size of the contract for tunneling these days.

Figure 15.3 shows the relation between risk sharing and contract types. The types of contracts could be turnkey, lump sum with fixed price, lump sum with price escalation, measurement of items, target amount, and cost reimbursement. Each of these types has risks shared between the owner and the contractor complementary to each other [1]. Another type of modern contract is BOT (build, operate, and transfer to owner).

The features of minimizing project cost [17] are shown in Figure 15.4. The total cost of the tunnel is a function of the economic factors and risk sharing. If the investigations are thorough, geological uncertainty is reduced as a result of the investigations, risks are shared by the owner and the contractor

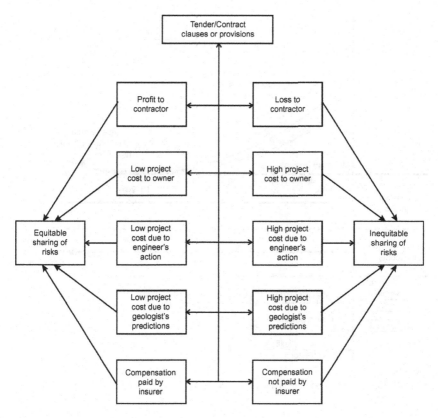

FIGURE 15.2   Conceptual model of risk sharing [15].

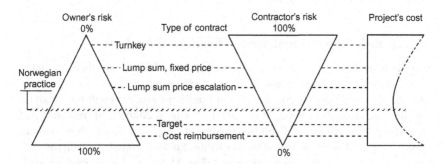

FIGURE 15.3   Risk sharing and contract types [1,16].

equitably, and, if the contractor is qualified, then the project cost can be minimized.

In a court of law, denial of contract is not valid without detailed reasoning. Further, engineers owe contractors and subcontractors an independent

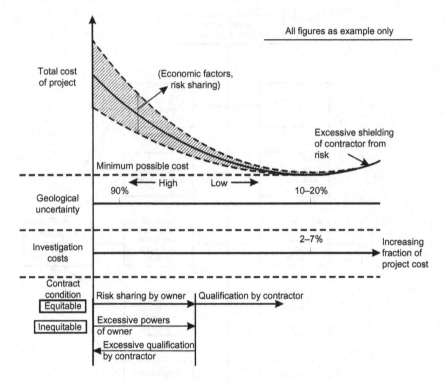

FIGURE 15.4 Features of minimizing project cost [17].

professional duty of care in the preparation of plans and specifications (ASCE, Civil Engineering, Vol. 73, No. 12, 2003).

15.2.1 Risk Management Tools—Fault Tree Analysis

Care is taken in the planning of an underground infrastructure that calculated risk during construction is not high as it creates tensions among engineers, but risk is normal during construction in fragile mountains. Fault tree analysis can be used to analyze a single or combined causal connection (relation) that precedes a negative event. Fault tree analysis is utilized either with or without quantifying probabilities for events. By using this tool, complex problems with many interacting events can be structured (Fig. 15.5). For further reading, refer to [18] and [19].

In tunneling, the best approach is a strategic approach for the management of risk and reduction of cost and time overruns, especially in complex geological conditions. As a result, tunnel design is basically a decision analysis problem of uncertainty management. The aim of risk management is not only to do things right, but also to do right things.

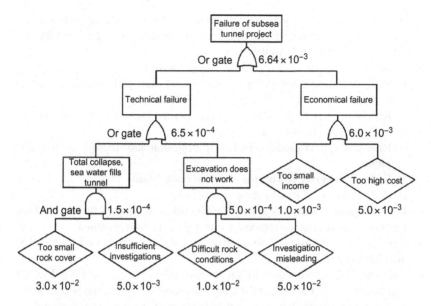

FIGURE 15.5 Example of a fault tree with "and gates" and "or gates" and evaluated probabilities for an undersea tunnel [20].

15.2.2 Recommendations of International Tunnelling Association [20]

- Tunnel failures have been the result of various reasons, such as insufficient site investigation, inadequate evaluation of risk at the planning stage, project understaffing, and mistakes during construction and operation phases.
- Provision may also be made for revised risk management plan and time schedule as agreed between the parties when the initial plan fails. Be prepared for the worst and hope for the best.
- Due regard should be taken for common clauses for hazardous events, such as
  - Complexity and maturity of the applied technology
  - Adverse unexpected ground and groundwater conditions (or geological surprises)
  - Subsidence and blasting vibrations on ground surface damaging structures and foundations
  - Technical and/or managerial incompetence
  - Human factors and/or human errors
  - Lack of sufficient communication and coordination between internal and external interfaces (workers)
  - Combinations of several unwanted events that individually are not necessarily critical
  - Lack of awareness causing risks (mistakes) of project

- It should be stated in the tender documents that the contractor is responsible for effective risk management, regardless of the extent and details of the risk information deriving from the owner.
- Types of risk covered in a contract are as follows:
  - Risk to the health and safety of workers and third-party people, including personal injury and, in the extreme, loss of life
  - Risk to third-party property, especially normal buildings, cultural heritage buildings, and infrastructure
  - Risk to the environment, including pollution and damage to flora and fauna
  - Risk of employer (owner) in delay to completion
  - Risk of financial loss to the employer (owner)
- Risk mitigation measures should be identified as long as the costs of the measures are not disproportionate with risk reduction obtained.
- Tenderer may be allowed to modify his bid (technically and financially) after opening of tenders.
- The tender offering lowest cost, on the basis of sum of bid price + risk cost + upgradation of technology of firm, and other costs (cost of delay + cost of litigation + cost of side effects, etc.), may be accepted by the owner from among a preselected list of prequalified firms.
- Information and training should be given as necessary to all personnel throughout the duration of a tunneling project. Rock mechanics experts should be employed in the organization.

15.2.3 Recommendations of International Standard Organisation [21]

The International Standard Organisation (ISO) has prepared several codes on risk management for petroleum industries, safety of machines, and fire hazards. ISO 31000 summarizes the principle of risk management as follows [21].

- Risk management is dynamic, iterative, and responsive to change (trial-and error-method).
- Risk management and its ongoing effectiveness require strong and sustained commitment by management of the organization as well as strategic and rigorous planning.
- In particular, risk management should be embedded into the policy development, business and strategic planning, and change management processes.
- Organization should hold information and training sessions.
- A consultative team approach is useful to
  - Bring different areas of expertise together for analyzing risks
  - Help ensure that risks are identified adequately
  - Ensure that different views are respected in evaluating risks

- The executive should define sharing of responsibilities for the risk management process and the level at which risk is high and requires treatment.
- Risk analysis is about developing an understanding of the risk.
- It is also important to consider interdependence of different risks and their sources.
- Results of monitoring and review should be recorded and published internally or externally.
- Risk management activities should be traceable. Records provide the foundations for improvement in methods and tools as well as the overall process.

15.2.4 Role of Engineering Leaders

The importance of leadership abilities in engineering is now realized. Strong engineering leaders became inspiring as they have made vast services to the people. *Experiences teach us that great persons are hidden sources of courage to their colleagues.* Strong engineering leaders are able to create interest in a project among engineers. The good leader can turn average workers into the top performing team of workers. Thus young engineers and scientists should request strong engineering leaders to solve their problems of underground infrastructure, risk management, and develop financial resources; and inspire courage and wisdom in all and leads us to the spectacular organized success. It is a pleasure to work with leading engineers and scientists. Strong engineering leaders should be given major powers and full freedom by the lawmakers of their countries. Thus efforts should be made to appoint strong engineering leaders as the chairperson of the vast civil engineering projects, if feasible. Success of the Delhi metro is one such example of strong engineering leadership in Mr. E. Sreedharan. He inspires that technical competence, punctuality, integrity and good moral values, and good health are the four pillars of organizational success. His engineers developed a sage-like peace even in the most trying situations. Leadership abilities of the strong leaders are evolving constantly with time. A recent example is the world leader Mahatma Gandhi, according to Albert Einstein. Because outdoor sports generate a large number of abilities to manage high risks, sport-loving nations are on the top. Sport-neglecting nations are a flop.

15.3 CONSTRUCTION PLANNING AND RISK

The following key questions should be asked during risk assessment [22].

- What can cause harm? (hazard identification)
- How often? (frequency assessment)
- What can go wrong and how bad? (consequence assessment)
- What is the likelihood of damage? (risk calculation)
- So what? (risk acceptability)

- What should be done? (risk management)
- Is risk insured? (risk insurance)

Owners should eliminate a known risk rather than try to transfer it. Active pre-contract construction planning would eliminate construction hurdles before they become sources of construction delays and disputes. This aspect is better done by the owner who has more time and is in a better negotiating position. By allocating the risk of negotiating all construction permits to the contractor, the owner would convert risk into a certainty, rendering the negotiations more hurried, less effective, and more costly than if he himself had done the homework before calling tenders. Vagueness in tender statements leads to all kinds of disputes. The ITA Working Group on Health and Safety in Work published "Guidelines for Good Tunnelling Practice" in November 1985. Tunnel engineers may find it useful practice.

15.4 TIME AND COST ESTIMATES

Usually, long tunnel projects take 5–10 years in complex geological conditions from the start of conceptual design until delivery of the scheme to the owners. A casual observer may consider it too long a period, but they are, in most cases, quite short, accounting for the complexities of the project. In early optimistic days of the project, the owner must make estimates of time and cost stretching over many years, but actually based on little solid information. The early estimates are publicized and become frozen. Any subsequent changes, even though based on more accurate data available later, are suspect in the public eye and result in a loss of reputation of the engineer and his profession. The time of completion is affected by confusion in risk assessment. There is a penalty clause for a delay in the completion of projects and a reward for early completion of the same. Owners assess penalties for late completion, but contractors inflate their bids for unreasonable schedules and fight back through the courts for extra payments, much to the detriment of the owner. Thus, sufficient time should be allotted for long tunneling projects, after careful thought and based on the construction times of similar completed projects. While imposing a penalty on the contractor for late completion, the opposite should be included, that is, payment of a bonus for early completion at a still higher rate to serve as an incentive. Delays and costs due to the owner's decision and approval processes and for his changeover on which the owner himself may have little control should be allowed in the contract. Means for providing necessary reimbursement and time and for reducing or eliminating costly standby time should be found. The cost of prevention of loss and risk is far more than the cost of loss. The contractor should recognize them and provide measures for equitable risk sharing without including such risk factors in his bid.

The owner should work out the cost of time that is revenue earned per day on completion of the project. Financial incentives to a contractor for an early completion of a tunnel should be proportional to the cost of time. Rules should not be rigid but flexible and should be humanitarian. Failure is not a punishable offence. Justice should be to the satisfaction of all (Mahatma Gandhi). Let us work for the glory of humanity.

"Better risk management leads to a better rate of tunneling."

REFERENCES

[1] Kuesel TR. Allocation of risks. Proc RETC 1979;2:1713–24.
[2] NAS. Recommendations on better contracting for underground construction. Washington (DC): National Research Council; 1976. p. 151.
[3] Tillman EA. Better contracting and better management for major underground construction projects. Proc RETC 1981;2:1563–74.
[4] Bhat HS. Contracting practices for tunnelling. Workshop on rock mechanics problems of tunnelling and mine roadways, Srinagar; 1986. p. 6.4.0–6.4.14.
[5] Muirwood AM, Sauer G. Efficacy and equity of the management of large underground projects. Proc RETC 1981;2:1032–44.
[6] Ribakoff S. European vs US construction contracting practices. Proc RETC 1981;2:1575–83.
[7] Samelson NM, Borcherding JD. Motivating foremen on large construction projects. ASCE JOCD 1980;106(1):29–36.
[8] Paulson BC, Aki T. Construction management in Japan. ASCE JOCD 1980;106(Co3):281–96.
[9] Duddeck H. Risk assessment and risk sharing in tunnelling. Tunnel Undergr Space Technol 1987;2(3):315–7.
[10] Klevian E, Aas G. Norwegian practice of risk sharing in tunnelling contracts, underground hydropower plants. In: Proceedings of the international conference on hydropower, Tapir, vol. 1. Oslo (Norway); 1987. p. 65–73.
[11] Badrinath HS, Verma M, Singh B. Effect of management conditions in planning tunnel construction. CBIP, international symposium on tunnelling for water resources and power projects. New Delhi; 1988. p. 53–7.
[12] Badrinath HS, Verma M, Singh B. Learning and experience curves as a management tool in tunnel construction. In: Proceedings of the national seminar on productivity in construction industry. MITS, Gwalior, India; 1989. p. 82–98.
[13] Egbert JS. Construction management of Washington metro. Proc RETC 1981;2:1612–26.
[14] Singh B, Gupta I. Wonderful law of negative entropy of healthy beings: applications. Conference on materials components and applications, Kalinga Institute of Industrial Technology, Bhubneshwar; 2003.
[15] Badrinath HS. Development of expert system for contractual risk sharing in tunnel construction. Roorkee (India): W.R.D.T.C., IIT; 1991. p. 183.
[16] Barton N, Grimstad E, Aas A, Opsahl OA, Bakken A, Pederson L, et al. Norwegian method of tunnelling. World tunnelling, Aspermont, UK, June and August 1992, UK; 1992.
[17] Sutcliffe H. Owner-engineer-contractor-relationship in tunnelling: an engineer's point of view. Proc RETC 1972;1:815–28.
[18] Sturk R. Engineering geological information: its value and impact on tunnelling. Doctoral thesis, Royal Institute of Technology, Stockholm; 1998.
[19] Ang AHS, Tang WH. Probability concepts in engineering planning and design. Vol II- Decision, risk and reliability, John Wiley, New York, 1984.

[20] Eskesen SD, Tengborg P, Kampmann J, Veicherts TH. Guidelines for tunnelling risk management: international tunnelling association, working group No. 2. Tunnel Undergr Space Technol 2004;19:217–37.

[21] ISO 31000. Risk management: principles and guidelines on implementation. International Organization for Standardization, Geneva, 2008. p. 18.

[22] Ho K, Leroi E, Roberds B. Quantitative risk assessment: application, myths and future direction. GeoEng 2000, an International conference on geotechnical and geological engineering, Melbourne, Australia. Technomic Publishing, Lancaster, vol. 1; 2000. p. 269–312.

Questionnaire administered to underground workers and aboveground workers in Japan by the National Land Policy Institute's Committee on Utilization of Underground Space [1]

> ## Questionnaire about utilization of underground space

Please answer as follows

Underground workers   Aboveground workers

Q.1 → Q.3 → Q.4   Q.2 → Q.3 → Q.4

Q.1 For worker underground

How do you feel about the underground space where you are now working? Please mark below.

A. Image of underground

	negative				positive	
	−2	−1	0	1	2	
dark						bright
noisy						silent
hot						cool
narrow						wide
monotony						variety
closed in						open
uneasiness						easiness
inconvenient						convenient
discomfort						comfort
danger						safety
unhealthy						healthy

B. Prevention of disasters and safety
1. I always feel uneasy
2. I sometimes feel uneasy
3. I don't feel uneasy at all
4. I feel ....

325

C. Indoor environment

1. My health is affected by an artificial controlled environment (air, light, noise, etc.) If you agree, please explain concisely

....

....

2. It is possible that my health is affected
3. My health is not affected
4. I feel ....

D. Psychological effects

1. The closed-in space, lacking outside views and natural lighting, creates heavy psychological pressure
2. I sometimes feel psychological pressure
3. I do not feel psychological pressure at all
4. I feel ....

E. What troubles do you have in underground space? Please mark ranking (1, 2, 3, ...)

☐ comfortable lighting and brightness
☐ prevention of disasters and safety
☐ healthy air conditioning and noise
☐ psychological effects
☐ ....

F. Working in underground space

1. I want to continue to work underground
2. I do not want to work underground
3. I may want to work underground
4. I...

Q.2 For aboveground workers (including skyscrapers)

How do you think you would feel if you worked underground?

A. Image of underground

	negative				positive	
	−2	−1	0	1	2	
dark						bright
noisy						silent
hot						cool
narrow						wide
monotony						variety
closed in						open
uneasiness						easiness
inconvenient						convenient
discomfort						comfort
danger						safety
unhealthy						healthy

B. Prevention of disasters and safety
1. Remarkable difficulty in preventing disasters and safety in comparison with a skyscraper
2. Difficulty in preventing disasters and safety is the same as for a skyscraper
3. Difficulty in preventing disasters and safety is less in comparison with a skyscraper

C. Indoor environment
1. It is not desirable to work in a completely artificially controlled environment (view, air, lighting, noise, etc.) in comparison with a skyscraper
2. It makes no great difference either way
3. It is desirable to work in a completely artificially controlled environment
4. It is ....

D. Psychological effects
1. I would feel heavy psychological pressure working in a closed-in space having neither outside views nor natural lighting.
2. I would sometimes feel psychological pressure
3. I would not feel psychological pressure at all
4. I would ....

E. If you worked underground, what troubles would you have? Please mark ranking (1, 2, 3, ...)
☐ comfortable lighting and brightness
☐ prevention of disasters and safety
☐ healthy air conditioning and noise
☐ psychological effects
☐ congregation of vagabonds
☐ ....

F. Working in underground space
1. I would want to continue working in underground space
2. I would not want to work in underground space
3. I may want to work in underground space
4. I ....

Q.3 For both aboveground and underground workers

Please mark and write about the following.

A. Type of occupation
1. desk work   2. field work

B. Type of industry
1. manufacturing; 2. construction; 3. wholesale and retail sale; 4. finance and insurance; 5. real estate; 6. transportation and communication; 7. electricity, gas, water service; 8. services; 9. official business; 10. other (   )

C. Sex
    1. Male   **2.** Female

D. Age
    ………. years old

E. Years of working
    ……….. in total
    ……….. years in present service

F. Place of work
    Name and address:

Floors: ……… floor above/underground of a building with ……… floors

Only for worker underground

A. Place of work
    1. subway (a. platform; b. office; c. operation; d. other)
    2. underground market; **3.** underground parking; **4.** underground factory;
    5. underground library; **6.** underground office; **7.** other (   )

Q.4 Please express any opinion about utilization of underground space
     for markets, offices, and so on or any requests to public administration
     regarding these matters

....
....
....
....
....

Thank you very much for your cooperation

REFERENCE

[1] Nishi J, Kamo F, Ozawa K. Rational use of urban underground space for surface and subsurface
activities in Japan. Tunnel Undergr Space Technol 1990;5(1/2):23–31.

Index

Page numbers in *italics* indicate figures and tables.

Printed in the United States
By Bookmasters